Detection and Classification of Ice

ELECTRONIC & ELECTRICAL ENGINEERING RESEARCH STUDIES

RADAR AND SONAR SERIES

Series Editor: **Dr. John Clarke,** *Royal Signals and Radar Establishment Malvern, England*

1. Detection and Classification of Ice
   Edward O. Lewis, Brian W. Currie *and* **Simon Haykin**

Detection and Classification of Ice

Edward O. Lewis
Department of Fisheries and Oceans, Burlington, Canada
Brian W. Currie
McMaster University, Hamilton, Canada
Simon Haykin
McMaster University, Hamilton, Canada

RESEARCH STUDIES PRESS LTD.
Letchworth, Hertfordshire, England
JOHN WILEY & SONS INC.
New York · Chichester · Toronto · Brisbane · Singapore

RESEARCH STUDIES PRESS LTD.
58B Station Road, Letchworth, Herts. SG6 3BE, England

Marketing and Distribution:

Australia, New Zealand, South-east Asia:
Jacaranda-Wiley Ltd., Jacaranda Press
JOHN WILEY & SONS INC.
GPO Box 859, Brisbane, Queensland 4001, Australia

Canada:
JOHN WILEY & SONS CANADA LIMITED
22 Worcester Road, Rexdale, Ontario, Canada

Europe, Africa:
JOHN WILEY & SONS LIMITED
Baffins Lane, Chichester, West Sussex, England

North and South America and the rest of the world:
JOHN WILEY & SONS INC.
605 Third Avenue, New York, NY 10158, USA

Library of Congress Cataloging-in-Publication Data

Lewis, Edward, O., 1947
Detection and classification of ice.
(Electronic & electrical engineering research studies. Radar and sonar series; 1)
Bibliography: p.
Includes index.
1. Sea ice. 2. Ice navigation. 3. Radar in navigation. I. Currie, Brian W. II. Haykin, Simon S., 1931– . III. Title. IV. Series.
GB2405.L48 1987 623.88′8 87-12687
ISBN 0 461 91641 2 (Wiley)

British Library Cataloguing in Publication Data

Lewis, Edward O.
Detection and classification of ice.—
(Electronic & electrical engineering research studies. Radar and sonar series).
1. Radar in glaciology
I. Title II. Currie, Brian W.
III. Haykin S. S. IV. Series
551.3′1 GB2401.72.R32

ISBN 0 86380 058 0

ISBN 0 86380 058 0 (Research Studies Press Ltd.)
ISBN 0 471 91641 2 (John Wiley & Sons Inc.)

Printed in Great Britain by Galliards (Printers) Ltd., Great Yarmouth

Table of Contents

Page

Page

Editor's Preface

This first book in the Radar and Sonar series is well worth my commendation. It covers a solid programme of research into radar for detection and classification of ice. The structure of the volume is archetypal for a Research Study with state-of-the-art review, theoretical considerations, and new experimental results. The subject attracts a high interest currently as it is very relevant to drilling operations, ice breaking, transportation, and navigation. The research is directly relevant to those working in the field of ice monitoring, but the approach taken and the coverage will prove valuable to other workers in the field of radar systems. The recommendations on selection of fundamental design parameters are particularly valuable.

John Clarke
RSRE, Malvern, U.K.

March 1987

Preface

Although much has been written on the surveillance of ice from the air and from space, little previous work has been done to improve the performance of a radar operating from a surface-based platform, e.g., a ship or a drilling rig. This type of radar, however, is the one which is used continuously, in all light and all weather, and represents the last line of defence. Present radars do not provide adequate performance, and improvements are required. This book addresses this gap in the knowledge by presenting the important elements of radar theory and ice properties, then reporting on a six-year research program conducted by the authors to systematically evaluate the radar design parameters that yield improved performance for surface-based radar surveillance of ice at sea. The book concludes with a description of ship trials in which a prototype of a new ice radar design implemented by the authors was successfully demonstrated.

Material concerning the separate studies of ice properties and surface-based radar theory is normally found only in separate volumes. This book ties these two studies together in a unique fashion, thereby providing a singular reference on the subject of ice-radar surveillance. The reader is assumed to have limited knowledge in one area of study or the other, or both. As such, the material is presented in a tutorial manner. This part of the book will be of interest to those beginning a study of radar and ice, or to non-technical personnel who need a knowledge of the issues involved in ice-radar operation.

The book then continues with a more detailed discussion of these subject areas, at a level suitable for reference by personnel working more actively in the field. For the engineer or operator involved in the surface-based detection or classification of ice at sea, this book provides a synthesis of data and information from a wide range of sources on most aspects of the ice-sensing problem. The book also contains an extensive bibliography of papers addressing this subject.

The most important contribution of the book, however, is the presentation and analysis of extensive experimental data gathered over the past six years, culminating in the successful demonstration of a functional system. Where appropriate, the theory and background material presented earlier in the book are combined with the experimental work to substantiate the theory and explain the observed results.

The first chapter of the book outlines the background to the research program and discusses the limitations of existing surface-based radar systems vis-à-vis the ice sensing problem. In Chapter 2 the history of radar for navigation in ice is described. Chapter 3 describes the physical and electrical characteristics of each of the ice types of interest, and their relationship to surface-based sensing of ice. The ice types of interest are first-year and multiyear sea ice, and ice of glacial origin (icebergs). Chapter 4 discusses radar theory as it pertains to this application. In Chapter 5, the experimental program, which concentrated on radar frequency and polarization effects, is described. Chapter 6 discusses a number of environmental effects as they apply to the problem of ice detection and classification. These effects include multipath propagation, atmospheric attenuation, anomalous propagation, and sea clutter. Chapter 7 discusses radar frequency considerations and the analysis of the radar frequency-related experiments. In Chapter 8, polarization is discussed, and the results of the polarization experiments are described. Chapter 9 summarizes the frequency and polarization results and proposes an explanation for the characteristics of back-scatter from ice. In Chapter 10, the effects of other radar parameters such as antenna height, azimuthal beam-width and transmitted pulse length are discussed. An important element in a new ice-radar design is improved signal processing and display techniques. Chapter 11 discusses techniques for improving target detection. These include scan-to-scan and pulse-to-pulse integration and the use of CFAR processors. Chapter 12 describes display improvements and image processing techniques. In Chapter 13, the configurations for ice classification and ice detection systems are given. Finally, in Chapter 14, results of recent shipboard trials of a prototype ice radar system are reported, and a summary of future work to be undertaken is presented.

The results described in Chapter 14 are very timely, reflecting experimental work carried out during November and December of 1986. The authors believe that this book represents the current state of knowledge on surface-based remote sensing of ice.

Acknowledgement

A research program of the magnitude described in this book requires the dedicated effort of many people, not just the ones whose names eventually appear on the cover. In particular, the authors would like to acknowledge the excellent technical contribution of Trevor Dyas and Bill Montgomery. We also acknowledge the work of Dianne Crabtree in typing the manuscript, Jean Fiddes in proofreading, and Lyn Arsenault in technical editing.

The funding for the experimental program was provided by the Canadian Energy Research and Development Fund without which the program would not have been possible. In addition, part of the involvement of McMaster University personnel was made possible by grants from the Natural Science and Engineering Research Council. The first author also wishes to acknowledge the encouragement and support received from senior personnel of the Department of Fisheries and Oceans.

We would like to thank those who have given permission to reproduce copyright material, and have acknowledged the permissions throughout the book.

We would also especially like to thank Canarctic Shipping Limited for the provision of ship time for testing of the prototype system.

On a personal note, the first two authors wish to acknowledge the support and understanding of their wives and families, particularly during long absences from home conducting the experimental program.

List of Abbreviations

ACP	azimuth change pulse
A/D	analogue to digital
ARE	Arctic Research Establishment
CA	cell averaging
CAGO	cell averaging greatest of
CFAR	constant false alarm rate
cm	centimetre
CRC	Communications Research Centre
CRL	Communications Research Laboratory
CW	continuous wave
dB	decibel
dBm	decibel referenced to 1 mw
EM	electromagnetic
FTC	fast time constant
GHz	gigahertz
HH	horizontal transmit/horizontal receive
HV	horizontal transmit/vertical receive
IF	intermediate frequency
IR	infrared
km	kilometre
kw	kilowatt
m	metre
MDS	minimum detectable signal
MHz	megahertz

mm	millimetre
MY	multiyear
nm	nautical mile
NRL	Naval Research Laboratory
P_d	probability of detection
P_{fa}	probability of false alarm
pdf	probability density function
PPI	plan position indicator
PRF	pulse repetition frequency
RCS	radar cross-section
RF	radio frequency
ROM	read only memory
SAR	synthetic aperture radar
S/I	signal-to-interference ratio
SLAR	side-looking airborne radar
SNR	signal-to-noise ratio
STC	sensitivity time control
T/C	target-to-clutter ratio
VDAS	video data acquisition system
VH	vertical transmit/horizontal receive
VHF	very high frequency
VV	vertical transmit/vertical receive

CHAPTER 1
Introduction

1.1 THE PROBLEM

With the worsening of the world oil crisis which began about 1973, Canada has taken steps to identify and develop frontier energy sources. In particular, extensive exploration and development work has been carried out in the Beaufort Sea, throughout the Arctic islands, and off the Canadian east coast. These areas are shown in Figure 1.1. In addition to energy exploration, there has also been exploration for, and development of, mineral resources in frontier areas. There are two active base metal mines north of the Arctic Circle, at locations also shown in Figure 1.1. The Beaufort Sea exploration, arctic mines development, and east coast explorations have all resulted in substantially increased shipping activity over the past ten years, and continued expansion is anticipated. Although the Arctic Pilot Project (proposed to carry liquid natural gas from Melville Island) has been delayed indefinitely, one shipload of crude oil was transported from the Arctic archipelago (Cameron Island) in 1985 and one in 1986.

In the past few years, hydrocarbon exploration off the east coast of Canada has become more active than in arctic waters. Major discoveries of oil at Hibernia and

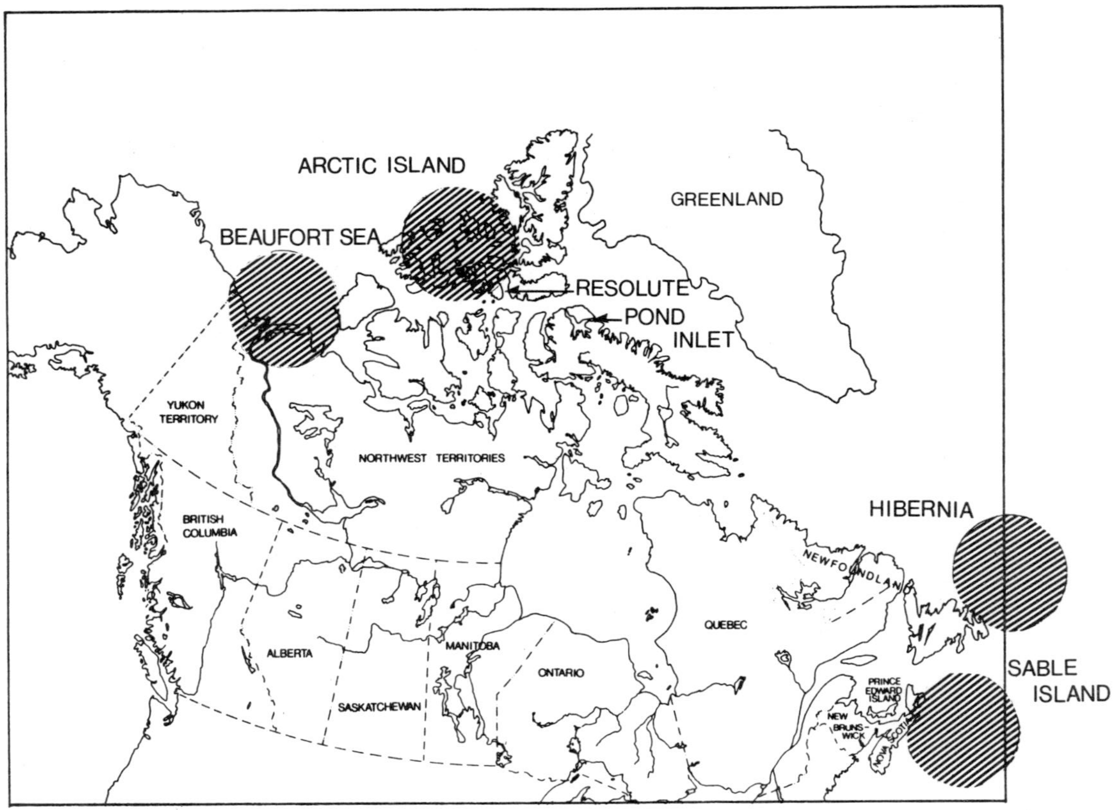

Figure 1.1 Areas of offshore development in Canada.

gas around Sable Island have accelerated the pace of east coast exploration.

1.2 ICE CONDITIONS

All the Canadian frontier developments have at least one element in common: they involve operations in ice-covered or ice-infested waters. The structure, type of ice, and risk to offshore development for each of the areas, however, are different. In the Canadian offshore environment, there are three distinct types of ice: first-year, multiyear, and glacier. The properties and characteristics of these ice types are described in detail in Chapter 3.

Table 1.1 summarizes the ice conditions encountered in Canadian frontier areas, their threat to offshore

Table 1.1 Ice conditions in Canadian frontier areas.

Ice feature	Threat	Ability to detect and classify	Area likely to occur
First-year ice	Low to moderate	Fair	All
First-year ice ridges	Low to moderate	Fair	All but east coast
Multiyear ice	Moderate to high	Poor	All, but rare off east coast
Multiyear ice ridges	Moderate to high	Fair	All but east coast
Ice islands	High	Fair	All but east coast
Icebergs	High	Good	East coast East Lancaster Sound
Bergy bits	High	Fair	East coast East Lancaster Sound
Growlers	Moderate to high	Poor	East coast East Lancaster Sound

operations, the ability to detect and classify, and the geographic area of occurrence. In this work, the term "detect" is used in the classical sense of "to spot or identify," while the term "classify" signifies "the ability to differentiate one form or type from another." Table 1.2 shows an abridged version of the World Meteorological Organization's definition of icebergs, bergy bits, and growlers. A list of definitions for some ice types is included in Appendix A.

Table 1.2 Definition of icebergs, bergy bits, and growlers. (See additional definitions in Appendix A.)

Iceberg	A massive piece of ice of greatly varying shape, more than 5 m above sea-level, which has broken away from a glacier, and which may be afloat or aground. Icebergs may be described as tabular, dome-shaped, sloping, pinnacled, weathered, or glacier bergs.
Bergy Bit	A large piece of floating glacier ice, generally showing less than 5 m above sea-level but more than 1 m and normally about 100 to 300 square metres in area.
Growler	Smaller piece of ice than a bergy bit or floeberg, often transparent but appearing green or almost black in colour, extending less than 1 m above sea surface and normally occupying an area of about 20 square metres.

In the Beaufort Sea, the ice environment consists of a mixture of first-year, and multiyear (pack) ice. During the summer months, the multiyear pack ice withdraws to about 300 to 400 km offshore, and the Beaufort Sea is essentially free of ice for about 90 days. However, by January, the fast (first-year) ice has grown out from the land to about the 20-m depth contour where it meets the incoming multiyear pack ice. The area of intersection of the fast and pack ice is known as the transition zone. This region is a dynamic area where heavy ridging and ice deformation take place. In the transition zone, there is a mixture of first-year and multiyear pack ice, as well as occasional fragments of ice islands. It is in this area that much of the Beaufort Sea exploration activity takes place. The major risk to offshore activities comes from the encroachment of the pack ice on to the drilling site. When production begins, there will be a further problem of navigation through an area of mixed first-year and multiyear ice, as well as the effect of ice on pipelines during the production phase. Multiyear ice poses a significant impediment to navigation by even a suitably equipped ice breaking vessel, and poses a real threat to the integrity of the ship.

In the Arctic islands, the main problem is encountered during transit through an ice cover with a large percentage of multiyear ice. Offshore drilling operations have a further problem related to ice motion.

All transportation routes to and from the high Arctic (including a potential route for oil transport from the Beaufort Sea) concentrate in the eastern end of Lancaster Sound. This confluence has highly dynamic ice conditions which, in addition to first-year and multiyear sea ice, usually has quite large numbers of icebergs, bergy bits, and growlers. During the summer months there is considerable (or even complete) open water as far westward as Resolute Bay; however, during the winter and spring months, a full ice cover can extend as far east as the entrance to Lancaster Sound. In addition to the problem of navigation through first-year, multiyear, and glacier ice, there is the problem of navigation through open water inhabited by bergy bits and growlers.

The ice situation off the east coast of Canada is limited to first-year pack ice, and icebergs, bergy bits, and growlers. Multiyear floes are infrequent off Labrador and rarely encountered off Newfoundland. The southern limit of the pack (first-year) ice rarely exceeds 45°N, with the median southern extent occurring at about 46°N. The pack, in general, does not go further eastward than about 45°W. Although the southern limit of iceberg occurrence is not clearly defined, it would appear that few icebergs pass south of 44°N and rarely are there reports of icebergs south of 40°N. The number of icebergs crossing 48°N (about the latitude of St. John's, Newfoundland) has been monitored over a long period of time; it has been found to vary from as few as zero in 1966 to over 2,000 in 1984. Both the pack ice and icebergs heavily influence activities on the Grand Banks and at the Hibernia exploration site (about 47°N). The operational situation off the east coast is complicated by the high sea states that can occur, with the resulting difficulty in detecting pieces of ice which would be hazardous to a ship or drill rig. A 500-ton growler, with freeboard of only 1 to 2 m and length of less than 10 m, poses a serious hazard to a ship under steam or to a stationary drill rig when the ice is accelerated by wave action. For the balance of this work, unless otherwise stated, the term iceberg will include growler and bergy bit.

1.3 MONITORING

From the foregoing, it can be seen that knowledge of both the presence and type of ice is required by vessels and drilling platforms. In a heavy ice cover, such as exists for much of the year in the Beaufort Sea, Arctic archipelago, or Northwest Passage, classification of ice type is of prime importance. In predominantly open water conditions, such as off Newfoundland, or Lancaster Sound and Baffin Bay in summer, detection of sea ice and glacial-origin ice is crucial.

Vessels and drilling platforms currently use conventional marine radars as their prime aid for monitoring ice conditions, and the dual requirement of detection and discrimination places demands on the radar system which standard radars are not designed to address.

To supplement the information supplied by marine radars, offshore operators are using a variety of other sensors, surface-based, airborne, or satellite-borne. For satellite and airborne platforms, the sensors include optical, infrared (IR), and passive and active microwave. For surface platforms, the sensors include IR, optical (including low-light-level), active microwave (radar), and sonar instruments.

1.4 OPERATIONAL REQUIREMENTS

The operational demands for ice information are generally broken down into categories based on the temporal and geographical limits of the ice information requirement. The terms used are: strategic, tactical, and close-tactical.

Strategic information usually would be required in the time frame of several days to several weeks or longer ahead of a planned activity, and would cover large geographical areas, such as the complete route of a ship navigating the Northwest Passage, or the complete upstream ocean environment for a drilling rig. This information could be available from historic sources, from satellite imagery, and from long-range patrol aircraft.

Ice information required for tactical purposes would cover the time period from a few hours to one or two days ahead, and cover a geographical area equivalent to the temporal requirement. In the case of a ship in transit, the geographical coverage would be the distance the ship would expect to cover in the following 12 to 48 hours. This information would be used for route planning for the tactical period. For a stationary offshore facility, the geographical coverage for tactical purposes would be that area from which ice could reach the facility within the tactical period. The ice information would be used for ice breaking, ice deflection, iceberg towing, or for preparing to move the facility. For a ship, this tactical information could be provided by other vessels upstream, or from aircraft overflying the route. For an offshore facility, the information could be provided by other vessels or facilities upstream or by aircraft.

For the close-tactical case, the information is required continuously for the immediate period and up to three to six hours ahead, covering the contiguous vicinity of the operation. This information can be provided by aircraft, if the aircraft is immediately and continuously in the vicinity, or if the ice conditions are stable enough that little or no change takes place from the time the ice is sensed by the aircraft and the time that the operation encounters the ice. Two airborne systems frequently used are side-looking airborne radar (SLAR), and synthetic aperture radar (SAR). Alternately, the information can be acquired from systems on board the vessel or drilling platform. Surface-based radar is still the main and most effective of these surface-based sensors.

The use of SLAR and SAR for ice detection and classification is well documented (Anderson, 1960; Johnson and Farmer, 1971; Kirby and Lowry, 1979; Mercer _et al_., 1980). SAR, in particular, can provide images of outstanding quality with the ability to differentiate ice types and, to a certain extent, to detect icebergs. The

MV Arctic, which is operated by Canarctic Shipping Company Limited, transports concentrated ore from mines on Little Cornwallis Island and Nanisivik on northern Baffin Island to European markets, and has made extensive and continuous use of SAR for navigational routing during the late spring and late fall periods.

From an operational point of view, the use of an airborne sensor for close-tactical requirements (that is the immediate vicinity of the platform and continuous availability) has several severe limitations. First, aircraft may not be able to overfly the ship or platform continuously. Second, although the sensors aboard the aircraft are all-weather, the aircraft must take off and land, which is strictly controlled by weather conditions. Under arctic conditions, a plane may be grounded at an airport (such as that at Resolute) for up to a week. Reliability, maintenance, and crewing requirements of the aircraft also must be taken into consideration. Third, the SLAR or SAR provides imagery at a discrete time, and any significant motion of the ice between the time the SAR is flown and the encounter of the ship with the ice will reduce the value of the radar imagery significantly. Finally, there are limits to the number of suitably-equipped aircraft; presently there are only five aircraft with SLAR or SAR systems on board available for site-specific support to an offshore operation in Canada.

For these reasons it is obvious that although airborne sensors provide unique, high quality ice information that is vital to offshore operations, the offshore operator must also have dedicated, full-time ice sensors aboard the vessel or platform.

1.5 SURFACE-BASED SENSORS

At the present time the surface-based sensor requirement is being met, albeit poorly, with existing (or slightly modified) off-the-shelf marine radar systems. These systems were not designed for the detection and classification of ice targets; rather, they were optimized for detecting and tracking large targets. In addition, very little design change has been made in marine radar systems in the last thirty years which improves its performance for these applications. As noted by the Encyclopaedia Britannica (1985),

> "Various innovations since the implementation of radar during World War II have not improved significantly the capabilities of radar to detect icebergs."

This statement can be expanded to include the detection and classification of ice in general.

All existing systems are single frequency, with the exception of one dual-frequency system made by Racal-Decca, even though (as will be shown in this work) at least two widely separated frequencies will improve the classification, and possibly the detection of ice targets. All existing systems use horizontal polarization only; however a significant advantage in both classification and detection by employing dual polarization will be demonstrated. Although these existing systems have relatively narrow azimuthal beamwidths, their pulse lengths are long (to achieve useful average power on target), resulting in large resolution cells. In the presence of sea clutter, this large resolution cell size results in poor signal-to-clutter ratios. Superior signal-to-clutter performance can be achieved by employing shorter effective pulses through the use of pulse compression.

A major deficiency of existing marine radar systems is the radar display itself. For a shipborne radar illuminating an ice cover, the range of signal intensities for targets of interest spans 30 to 40 dB or more, while the dynamic range of the conventional plan position indicator

(PPI) display is limited to less than 12 dB. When the display is adjusted to show small targets, larger targets cause saturation of the display. This situation conceals the difference between smaller, less hazardous targets and larger, dangerous targets (including icebergs), since they are all displayed at essentially the same intensity.

Another problem results from the spreading of the radar energy with range. Targets near the radar reflect more energy than similar targets further away, since more incident energy falls on them. For a surface-based radar illuminating a distributed target (the ice surface), this decrease in incident radar energy varies as $1/R^3$ and after the transition range (described later) as $1/R^7$, where R is the range to the target. If the display is adjusted to show targets close to the radar, similar-sized targets further away can easily be lost because of this range dependence. Conversely, if the display is adjusted to show long-range targets, strong blooming near the display centre results in the loss of individual short-range targets.

A lesser, though still important, limitation of conventional display systems results from the fact that the PPI image is maintained on the display by employing tube persistence. This requires that the display be viewed under reduced lighting, either through a viewing hood or by darkening the ship's bridge. During daylight hours it is extremely difficult for more than one person to view the display at one time. Further, since the variable range marker is highlighted only by the rotating sweep, range measurements cannot be made until the sweep updates the area of interest.

1.6 ICE/RADAR INTERACTION

A further significant impediment, particularly to the classification of ice types and features, is the limited understanding of the interaction of radar waves with sea ice at low grazing angles. Most research has concerned the interaction of radar waves with sea ice from aircraft- and satellite-borne radars, which involve much larger grazing angles when compared to surface-based radars. Nevertheless, Onstott _et al_. (1979) commented about all ice interaction understanding by stating that there was ".... a general lack of fundamental knowledge of the interaction of microwaves with the various types of sea ice." Researchers working on the low grazing angle (surface-based) problem are regularly hampered by a lack of low grazing angle radar-ice data.

1.7 PROGRAM OBJECTIVES

There were three main purposes addressed by the research described in this book:

a) To develop a better understanding of the interaction between waves transmitted by a surface-based radar and ice.
b) To determine the radar parameters best suited to classify and detect ice.
c) To develop improved methods of displaying the radar images, including the use of image processing techniques.

1.8 SUMMARY OF THE BOOK

The main thrust of this book is directed towards improved navigation in ice-infested and ice-covered waters through the use of a radar system optimized for the ice environment.

The book begins by reviewing the history of radar in ice navigation and discusses radar theory as it pertains to this application. The effect of multipath interference on the propagation of the radar wave is described and experimental results of propagation experiments over an ice cover are presented. Atmospheric data from the east coast of Canada is analyzed to predict propagation conditions over open water.

A chapter is devoted to the physical and electrical properties of the three major ice types: first-year, multiyear, and glacial ice. Methods in which these characteristics can be exploited to improve detection and classification of ice are discussed.

The major contribution of this book is the analysis of data collected during four research experiments. The critical parameters analyzed are polarization and frequency. It is shown that by employing a cross-polarized radar channel, significant improvements in differentiating ice types can be achieved. In addition, by using both the like- and cross-polarized radar returns, improvements in the detectability of ice targets in open water are possible. The radar returns from various ice types in an ice cover are shown to be frequency sensitive. By using lower frequency systems, general surface roughness can be determined, while the use of higher frequency systems enhances identification of hazardous multiyear ice.

Based on the experimental data, values of the normalized radar cross-section for the three ice types are derived. Using these values, the differences in the normalized radar cross-section can be used to develop a classification system capable of identifying ice which is threatening to a vessel.

The book discusses the use of image processing to improve the utility of ice radar images. By using a digital display and a few simple image processing techniques, the interpretability of ice radar images can be substantially improved.

The book concludes with a description of a recommended radar system for use in ice navigation and presents results of the evaluation of a prototype system in November and December of 1986.

The work described in this book concentrates on the evaluation and use of non-coherent radar systems for the detection and classification of ice. All existing civilian marine radar systems are non-coherent. However, on several occasions possible advantages to be gained by the use of coherent radars (that is, systems which process both amplitude and phase information) are discussed.

CHAPTER 2
History of Marine Radar for Navigation in Ice

2.1 EARLY RADAR HISTORY

The early history of radar in general has been well reviewed; Page (1962), Luse (1981), Skolnik (1985), IEE Proceedings - A, Special Issue on Historical Radar (1985). Only a general summary of some of the key elements of marine radar development (with particular relevance to ice navigation) will be presented here.

The first installation of radar for commercial civilian marine use occurred in 1935 with the installation of a microwave radar aboard the French liner Normandie. This radar did not possess a rotating antenna nor a plan position indicator. In 1935, Kuhnold demonstrated a 600-MHz pulse shipboard radar system to the German navy.

During 1936, the U.S. Naval Research Laboratory (NRL) developed the first radar duplexer which allowed the use of a single antenna for transmitting and receiving. This development solved one of the major impediments to installing effective radar systems aboard ship. The first shipboard radar (operating at 200 MHz) with a rotating antenna was tested aboard the USS Leary in April of 1937. The display for this radar was an A-scope (a display of amplitude versus range). It was during these tests that the first sea clutter returns were observed.

At the same time, sea trials were going on in Britain with a 7-m wavelength system.

A major breakthrough towards modern radar capability occurred in 1940 with the development of a 10-cm (S-band) cavity magnetron at the University of Birmingham, England. Then, in the spring of 1941, the first microwave radar with a plan position indicator (PPI) was installed on the destroyer USS Semmes. This radar was the predecessor of the modern marine radar system.

2.2 RADAR AND ICE

The first documented use of a marine radar in an ice environment is reported by Luse (1981). The incident occurred in the spring of 1943 when the U.S. Coast Guard icebreaker Storis was escorting a troop carrier through icefields off Greenland. Luse reported that with the radar gain turned down he could see the escorted vessel as well as a fjord entrance, while with the gain turned up he could see the leads through the ice field.

By the end of the Second World War, X-band (3 cm) marine radars employing PPI displays were in common use. Following the war, surplus military radars were installed on many commercial ships and many of these operated in ice-infested waters.

During 1945 and 1946, the U.S. Coast Guard carried out a program in the Grand Banks area of Newfoundland to investigate the detection of discrete "lumps" of ice. The program employed American service radars at X- and S-band with both PPI displays with the antennas scanning, and A-scope displays with the antenna fixed stationary on a target. This experiment resulted in the first quantitative analysis of radar returns from ice. The results from this experiment were classified and were not released until they were consolidated with work reported by Budinger et al. (1960), discussed later in this section.

The National Research Council of Canada released a report in 1947 entitled Canadian Marine Radar, which in the final section discussed radar in Arctic waters. The report briefly discusses trials with the Type 268 X-band radar (developed by the National Research Council and produced in quantities exceeding 1,600 during the war) in detecting icebergs and growlers in Arctic waters.

One of the first detailed reports of ice detection by shipborne radar was that of Larsson (1949). Larsson reported on tests of X- and S-band radars aboard the Swedish icebreaker Ymer in 1947 and 1948. Some of the observations from these experiments are listed below:

a) Smooth ice-surfaces did not show up on the screen, but reflections appeared from areas where the surface had been broken. The broken rugged ice-surfaces always gave clear reflections up to 2 to 3 nm (nautical miles) away, with the 3-cm (X-band) system giving a clearer picture.
b) The relative strength of ice returns could be obtained by adjusting the gain control to compensate for the limited dynamic range of the PPI.
c) Old channels in the ice covered over with snow and invisible to the naked eye were often visible on the screen because of the radar signal's ability to penetrate the snow.
d) A direct relationship was noted between the intensity of the returns on the radar screen and the resistance to navigation through the ice; with the brighter the return, the heavier the resistance.
e) Firm ice could be distinguished from open water by the lack of sea clutter.

Le Page and Milwright (1953) reported results of tests carried out on board the Canadian icebreaker N.B. McLean to investigate the detection of ice hazards in Hudson Bay and its approaches. As a result of experiments carried out during the voyage, Le Page concluded that "radar is an invaluable aid to ships navigating in bad visibility in ice-infested waters, provided it is used wisely and the limitations of the radar are known." Other observations made include:

a) In calm seas, icebergs could be detected out to 15 to 20 miles, while small growlers were detected at two miles.
b) Under rough sea conditions when sea clutter extends beyond one mile, growlers large enough to cause damage to the ship can go undetected.
c) Fields of pack ice should be detectable under all sea conditions out to three miles.

Perry (1953) reported similar investigations aboard the SS *North Anglia* on a voyage to Port Churchill. Perry's results were similar to those of Le Page above. In smooth seas, growlers could be detected at two to three miles, while in rougher seas growlers sufficiently large to cause disaster could not be detected. On one occasion, close pack ice looked like sea clutter on the one- and three-mile ranges. Hummocks in the ice could only be isolated using the anti-clutter control.

Hood (1954a; 1954b) analyzed reports for performance of marine radar in detecting ice submitted by commercial shipping entering Hudson Bay. He concluded that for icebergs, the ranges of detection were reasonably close to theoretical and experimental data using a $1/R^4$ range relationship.

With the possible exception of the work done in 1945 and 1946 off Newfoundland, the first major, systematic experiment conducted to quantify the capability of marine radar to detect ice of potential hazard to shipping was carried out by the U.S. Coast Guard between March and October, 1959. The results of these experiments are reported by Budinger *et al*. (1960). These results were based on detailed analysis of 128 ice targets from the 1959 experiment and 24 targets from the 1945 experiment. From the results of the experiment, the maximum range of detection for the 152 targets conformed very closely to the anticipated $1/R^4$ relationship. However, on several occasions it was not possible to detect growlers in sea clutter. The reflection coefficient for icebergs on the Grand Bank was found to be about 0.33, and icebergs were found to reflect radar energy 60 times less than a ship with an equal-sized physical cross-section. During the experiment, subnormal propagation conditions were encountered and Budinger noted that (during the spring) subnormal propagation conditions would be the rule rather than the exception. He commented that this would lead to slightly reduced ranges of detection for small targets and greatly reduced ranges for large targets. Budinger concluded with the following warning:

> "All shipmasters, mates, and owners are warned that safe passage through iceberg areas of the North Atlantic Ocean cannot be assured by the use of radar."

During 1966, Holden of Decca Radar, as reported by Williams (1973), carried out controlled trials on the detection of growlers in various sea states. Contrary to expected results, it was found that for the detection of growlers in high sea states a wider beamwidth, lower azimuthal resolution antenna, gave superior results to a narrow beamwidth antenna.

Tabata *et al*. (1967, 1969a, 1969b) and Tabata (1975) reported on work carried out on the detection of pack ice off Hokkaido, Japan, using three coastal C-band (5 cm) radar stations. During these experiments it was found that the pack ice could be tracked over extended periods of time. From the tracking of ice, drift velocities could be determined. Ice-free patches (polynyas) within the ice pack could be determined by their low returns, while pressure ridges could be identified by their bright returns. With the heights of the radars being between 220 and 400 m above sea level, pack ice could be observed out to 60 to 85 km.

Hagman et al. (1976) reported on results of marine radar studies carried out as a part of the Sea Ice-75 program. During this program, X- and S-band radars were operated from the icebreaker TOR lodged in an area of ice. Some of the conclusions they reached were:

a) The ship's radar could distinguish the difference between open water and ice, open water and the ice edge, ice ridges and level ice, and vast thick floes refrozen in level ice.
b) The X-band radar gave a clearer picture than the S-band radar.
c) With the heights of the antennas on the TOR at 22 to 24 m, this, in general, gave good information on the ice situation out to 1.5 nm. Larger ridges (>2 to 3 m) gave radar echoes out to 3.5 nm.
d) The ship's radar could not see the difference between:
    i) snow-covered ice and snow-free ice unless the snow-covered ice was very rough;
    ii) the height of the ice ridges;
    iii) the thickness of the ice; and
    iv) a narrow channel and a small ridge.

During the late 1970s and 1980s, considerable work was carried out by or for the Canadian Government and the Canadian offshore resource industry on the detection, classification, and tracking of ice. Benedict and Hall (1979) prepared a report examining the benefits of a multisensor system including radar, for the detection of icebergs. Subsequent to this report, the Canadian Government, in conjunction with the offshore resource industry, funded field trials to evaluate the various sensors. Unfortunately, the results of this research are proprietary. Beginning in 1981, the Canadian Department of Fisheries and Oceans, in collaboration with the Communications Research Laboratory of McMaster University, has undertaken a systematic study of surface-based radar classification and detection of ice. The results are the subject of this book and some of the results are reported in Lewis and Currie (1984a, 1984b), Lewis *et al*. (1985), Currie and Lewis (1985), and Haykin *et al*. (1985).

2.3 SUMMARY

This chapter reviewed the history of the use of radar in ice navigation and research. During the period from the first documented use of radar in an ice environment in 1943 as reported by Luse, until the early 1980s, all navigation in ice (at least for civilian applications) was accomplished using conventional, commercial marine radar equipment. It was only in the early 1980s that equipment was specially modified or constructed to operate in an ice environment. These modifications, however, have been on a limited scale, and, with the exception of combining X- and S-band radars, have concentrated exclusively on radar display improvements. In addition, until the early 1980s the experimental programs undertaken have been confined to using conventional marine radar hardware. It was not until the commencement in 1981 of the research discussed in this book that a serious evaluation and examination of all radar parameters was undertaken. This research has led to a better understanding of the interaction between a surface-transmitted radar signal and ice, and has allowed the specifications to be drawn for a ship-borne radar dedicated to ice sensing.

CHAPTER 3
Ice as a Material

Ice is a non-magnetic, moderately-conductive material which displays wide spatial and temporal variability. For the purposes of this research, the ice that is of interest is sea ice and glacier ice. The two types are substantially different in physical and electrical characteristics and completely different in origin.

3.1 SEA ICE

Sea ice is an anisotropic material with wide spatial and temporal variability. The properties of the ice are a function of its history, temperature, salinity, density, and age, amongst other factors.

The ice cover can be divided into three major categories: first-year, second-year, and multiyear. First-year ice has grown within the current ice year (late October to July or August) but does not survive the summer melt season. Second-year ice has survived one summer melt season and has continued to grow during the second ice season, while multiyear ice has survived several melt seasons.

Ice Growth

The formation of sea ice is a complex process that begins at -1.8°C in normal saline water of thirty-five parts per thousand (35°/₀₀). If the ice forms in calm water, there is relatively rapid horizontal growth because the ice crystals lie with their axis of most rapid growth (a-axis) horizontal, and the c-axis vertical. The growth process consists of the formation of minute spheres of pure ice which, with rapid growth, change into thin circular discs. These discs in turn form into hexagonal, dendritic stars. This process is shown in Figure 3.1. The change from disc to star signals the start of rapid growth (Kumai and Itagaki, 1953), with the result that the star-like crystals overlap and freeze together. The resulting surface is smooth with little horizontal variability and is termed "sheet ice" (Wilson et al., 1954).

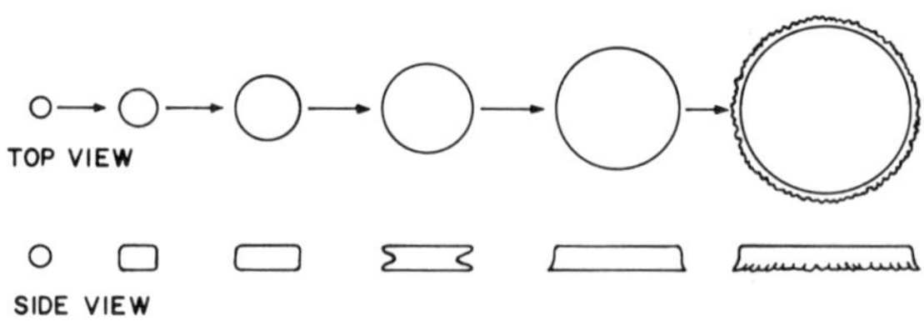

Fig. 3.1 Ice crystal growth process (Arawaka, 1954).

When the ice forms in more turbulent water the mixing process results in more crystals being formed per unit volume, which favours discordal rather than hexagonal stars. The higher density of crystals causes more abrasive action and results in the arms breaking off any star crystals that form. This randomly-oriented collection of ice crystals, called frazil, is driven by wind action into an unconsolidated, soupy opaque mass known as grease ice, which is only 20 to 40% ice by volume. Grease ice can be seen in Figure 3.2 between the open water at the lower part of the photograph and the consolidated ice at the top. The frazil ice can also form into more consolidated forms such as pancake ice, roughly circular pieces of ice with upturned edges as shown in Figure 3.3. The interaction between the pancake pieces pushes freshly formed frazil crystals upwards to form the raised edges.

Fig. 3.2 Grease ice in centre of photo occurring between open water at the bottom and consolidated ice at the top.

The sequence of ice formation in turbulent water is from open water containing frazil ice, to grease ice, to pancake ice plus grease ice, to joined pancakes, to a complete cover of pancake ice. If, during the process, the turbulence abates, grease ice will form between the pancakes, thus forming a complete ice cover as shown in Figure 3.4.

After this cover has formed, sheet ice will form below the pancake and grease ice. As a result of the general turbulence associated with the formation of the pancake and grease ice, this layer also consists of randomly

Fig. 3.3 Formation of pancake ice.

Fig. 3.4 Consolidation of ice cover.

oriented ice crystals. However, as the ice continues to grow below this layer, a preferred orientation of crystal growth develops. When the ocean surface becomes covered with ice, the growth rate is substantially reduced by the insulation of the seawater from the colder air. The growth rate of the ice is controlled thereafter by the temperature gradient in the ice cover.

The growth of ice crystals in a preferred orientation is a complex subject which is well developed in Weeks and Ackley (1982) and will not be repeated here. In summary, however, crystals which are aligned with their c-axis horizontal have the maximum rate of growth in the vertical direction. These fast-growing crystals in effect choke out other slower-growing crystals, resulting in an ice structure in which crystals with the c-axis aligned horizontally predominate (Perey and Pounder, 1958). This transition from random orientation to the preferred c-axis horizontal orientation takes place within a vertical distance of between 5 and 10 cm, although the transition may take place at up to 1 m from the surface in the case of a thick frazil layer. Not surprisingly, this layer is known as the transition zone. The ice below the transition zone is columnar in composition.

Because of the initial thinness of the transition zone (20 cm) and because "it sublimates away and recrystallizes into the snow cover" (Weeks and Ackley, 1982), a large part of the seasonal ice cover is columnar ice with c-axis orientation which is very close to horizontal. The thickness of this seasonal ice is typically between 1 and 2 m.

As the ice sheet grows, individual crystals project downward into the seawater. As more water freezes, much of the salt is expelled, increasing the salinity of the water at the ice/water interface. However, some of the brine is trapped within the growing ice, both within the crystal structure itself, but more predominantly in the interstitial zones between the crystals. The salt thus trapped is present both as liquid and solid inclusions. As the ice continues to grow, the ice crystals bridge together, trapping the liquid brine in elongated pockets (brine pockets), as shown in Figure 3.5. A brine pocket is a bubble of brine surrounded by ice, that has either limited or no communications with other brine masses (Niedrauer and Martin, 1979). The amount of brine trapped is a function of the salinity of the seawater and the freezing rate. The faster the freezing rate, the more brine that is entrapped. In frazil ice, the brine pockets are largely spherical (Campbell et al., 1978), while in columnar ice the brine pockets are elongated in the vertical direction.

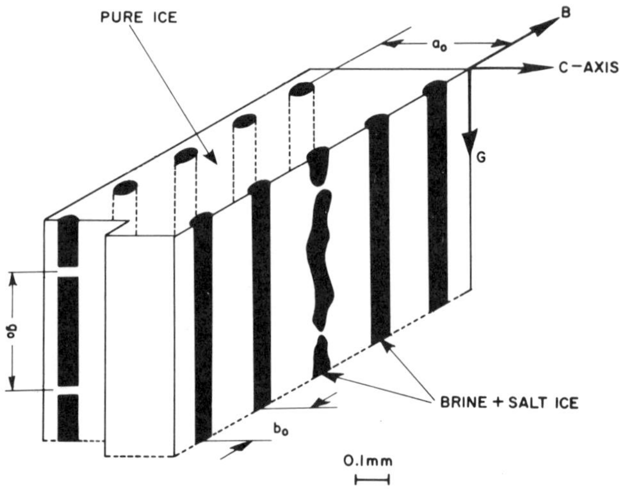

Fig. 3.5 Schematic diagram of brine inclusions in sea ice (Assur, 1958)

It is interesting to note, although not significant from a radar point of view, that the size of the ice crystals increases with increasing depth.

Addison (1969) notes that the diameter of the brine pockets is about 0.05 mm, while the length varies from 2 to 3 cm. The spacing between brine inclusions, the brine layer spacing a_0, is usually 0.5 to 0.6 mm, although there is a gradual increase in the spacing with depth. Rather than being directly related to depth, it appears that a_0 is related to the ice growth rate, which is itself generally related to depth. Results by Nakawo and Sinha (1981) show that a_0 is inversely proportional to growth rate. Thus slow growth rate produces large a_0 values.

As the ice grows and ages during the growth season, there is a continuing reduction in the entrapped salts by several mechanisms referred to as brine drainage. The mechanisms include brine pocket migration, brine expulsion, and gravity drainage. By far the most significant mechanism is gravity drainage, which is the only one that will be discussed here; Weeks and Ackley (1982) give a complete description of all mechanisms.

Under the force of gravity, the liquid brine drains through channels in the ice known as brine drainage channels. These channels are like the root system of a tree, but inverted, such that the smallest channels are near the surface, coalescing and widening with increasing depth in the ice. Near the bottom of the ice, the channels have a horizontal spacing of about 15 to 20 cm and a diameter of about 1 cm (Schwartz and Weeks, 1977). In addition to these larger channels, Lake and Lewis (1970) also reported smaller tubes near the ice/water interface, with diameters of 0.3 to 0.5 mm and a density of 42 tubes/cm^2.

As the temperature of the ice decreases, the liquid inclusions in the brine pockets begin to freeze and the salts begin to precipitate out, thus reducing the liquid brine volume. At -8.2°C, sodium sulphate decahydrate starts to precipitate out (Anderson, 1960), while at -22.9°C sodium chloride dehydrate begins to precipitate out, causing a large change in brine content. Other salts precipitate out at still lower temperatures, including potassium chloride at -36°C and calcium chloride hexahydrate at -54°C.

The ice surface, far from retaining the relatively smooth structure of original formation, undergoes significant and continual change. Under forcing from the winds and ocean currents, adjacent ice sheets impact under intense pressure. If the ice sheets are relatively thin, one sheet will ride up over the other; a phenomenon known as rafting. Rafting can occur repeatedly until several layers of ice are stacked one on top of the other. If the resulting ice layers remain level, the ice will become consolidated. Usually, however, the sheets will be haphazardly arranged both on and below the surface, and the gaps between the ice sheets below sea level will be filled with seawater. The density of ice and the force of gravity will cause more ice to be below the water line than above. Rafting takes place if the ice sheets are less than a metre or so in thickness.

If thicker sheets collide, the ice sheets buckle at the intersection of the ice floes or at some point of weakness in the ice sheet. This buckling produces pressure ridges, which for first-year ice can reach heights of 2 m or more. The pressure ridge is characterized not only by the above-water ice structure, which is called the sail, but by a larger below-water structure known as the keel. In multiyear ice, which will be discussed shortly, pressure ridge sails can reach heights of 10 m.

At the end of the growth season, the sea ice begins to ablate with warming air temperatures. In some areas, the ice melts completely and new ice forms at the beginning of the next ice year. However, in some areas the ice survives the melt season, becomes second-year ice in the fall and begins to grow thicker. With survival through a subsequent melt season the ice becomes multiyear ice. Multiyear ice usually grows in thickness to about 2 to 5 m, although thicknesses in excess of 10 m (Walker and Wadhams, 1979) have been reported. In general, surface melting will result in the ablation of the older, topmost section of multiyear floes such that the oldest ice would normally be 10 to 12 years old.

During the summer melt season, the remaining frazil ice, if any, and the top of the columnar ice melts, leaving mostly columnar ice exposed at the surface. This exposed ice undergoes extensive thawing and freezing cycles and recrystallization, with the result that the top layer of ice (about 1 m deep) changes from opaque to very hard, clear ice with a large number of bubbles. The process of brine drainage, as previously discussed, is substantially aided by flushing of fresh surface water through the ice during the melt season. Weeks and Ackley (1982) concluded that "it is believed that flushing is the most effective mechanism for desalination." It is during the period when flushing starts that the greatest decrease in salinity of the ice takes place.

During the summer months, the characteristics of both first-year and multiyear ice change substantially. As the temperature rises, the snow on the surface of the ice begins to melt, the snow thickness decreases, and the snow becomes very damp. During this period, the surface of the snow also can become increasingly rough.

Eventually, the snow on the relatively smooth first-year ice has all melted and substantial areas (75 to 95% in Holt et al., 1985) of the first-year ice are flooded with a layer of relatively salt-free water, several centimetres thick. On the surface of the multiyear ice, much of the snow has melted and, because of the ice's more undulating surface, the water has collected in melt pools which can be up to 20 cm deep (Gogineni et al., 1984). In addition, the surface of the higher ice features begins to melt. In the weeks that follow, some of this water drains off the surface through cracks or holes in the ice, while later in the season most of the water eventually drains through the ice structure itself. Thus, it can be seen that for the major part of the summer season, the surface of the ice is wet, flooded, or covered with damp snow.

In addition to contributing to reduced salinity in the ice column, the melt season causes substantial smoothing and rounding of surface features. Pressure ridges, rubble fields, and rafted zones are all rendered less angular. Figure 3.6 shows a typical multiyear floe with significant relief, but with well-rounded features. This same process causes pressure ridges to melt and then refreeze in a more consolidated form, resulting in the ridges becoming much stronger and more difficult to break.

The melting and thawing process produces pools of low-salinity water on the floe surface and the remaining water subsequently freezes, producing smooth areas of nearly freshwater ice. The structure of the resulting multiyear floe is shown diagrammatically in Figure 3.7.

Salinity

The salinity of sea ice is related directly to its rate of growth, its age, and the ice and atmospheric

Fig. 3.6 Multiyear ice floe with significant vertical relief.

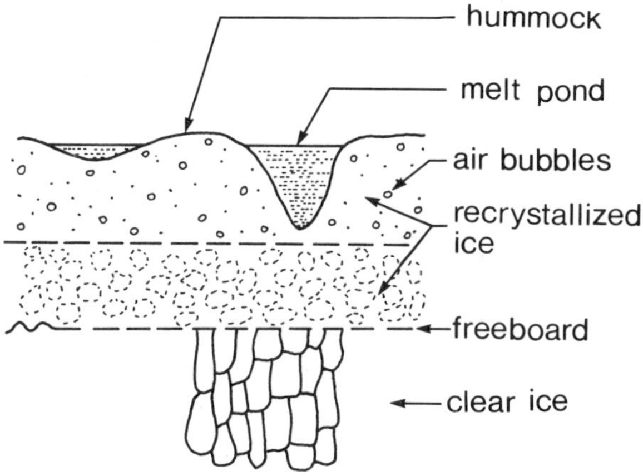

Fig. 3.7 Schematic diagram of multiyear ice floe structure (Vant et al., 1978). Reprinted by permission of the American Institute of Physics.

temperature. As discussed earlier, as the ice ages, brine drainage takes place and the ice becomes less saline. Although the salinity of seawater in the open ocean is in the order of 35°/₀₀ (parts per thousand), the initial salinity of the ice can be considerably higher.

A measure of the salinity of the ice is determined by taking a small sample of the ice and melting it, and measuring the salinity of the resulting water. Thus, the salinity is an average of the brine pockets and the surrounding relatively salt-free ice.

During formation, the salinity of the thin layer of new ice can be greater than 50°/₀₀ (Martin, 1979), although 25°/₀₀ seems a more usual value. There is an initial rapid decrease in salinity with time, but for cold arctic ice the salinity reaches a quasi-stable value which then decreases very slowly during the rest of the growing season. Nakawo and Sinha (1981) reported a decrease in salinity from 25°/₀₀ on 2 December to 8.5°/₀₀ on 9 December. Figure 3.8 shows this rapid initial decline as well as the slow decrease in salinity over the remaining period of the ice growth. Nakawo and Sinha (1981) noted that desalination increased rapidly once the warmer weather of May and June arrived.

In warmer conditions off Labrador, Weeks and Lee (1958) showed the same rapid decline in salinity shortly after ice formation (see Figure 3.9), but also showed a continuing decline in salinity. The warmer ice temperatures at this site lead to continuing desalination.

Typical salinity values for arctic first-year ice, after initial rapid desalination, appear to be about 10 to 20°/₀₀ near the ice surface, 4 to 6°/₀₀ in the bulk of the ice sheet, rising to 10 to 30°/₀₀ at the ice/water interface. Figure 3.10 shows a typical salinity versus depth plot for first-year ice.

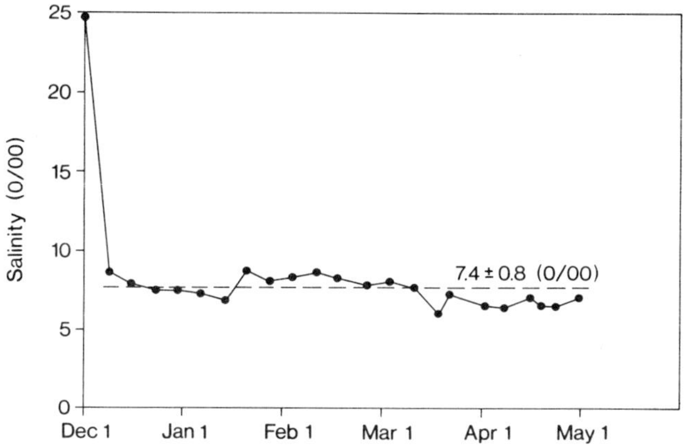

Fig. 3.8 Variation of salinity with time for northern first-year ice at 40-cm depth. Broken line indicates average salinity excluding initial high values (Nakawo and Sinha, 1981). Reproduced by courtesy of the International Glaciological Society from <u>Journal of Glaciology</u>, <u>27</u>(96), p. 315-329.

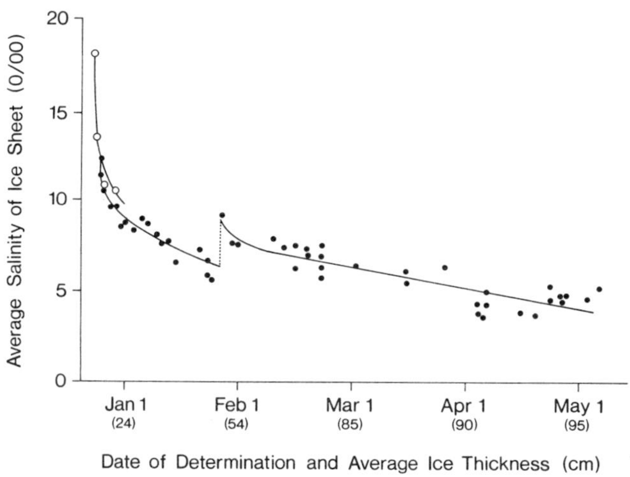

Fig. 3.9 Average salinity of ice sheet versus time for Labrador coast ice (Weeks and Lee, 1958). Reproduced by permission of the Arctic Institute of North America from <u>Arctic</u>, <u>11</u>(3), p. 135-156.

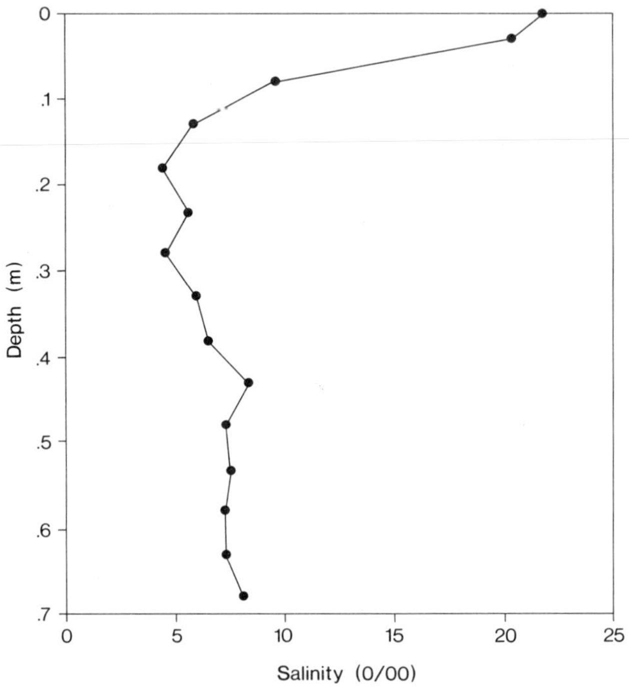

Fig. 3.10 Typical salinity versus depth profile for first-year arctic sea ice (Martin, 1979). Reproduced by courtesy of the International Glaciological Society from Journal of Glaciology, 22(88), p. 473-502.

As the ice goes through the melt season, significant brine drainage takes place. Figure 3.11 shows the reduction in salinity and change in the shape of the salinity profile during the rapid desalination period. Not only does the net salinity decrease, but there is a substantial reduction in salinity near the surface and near the ice/water interface.

Figure 3.12 shows an average profile of salinity for multiyear ice. The salinity is near zero at the ice surface, increasing to 2 to 3°/₀₀ in the bulk. These

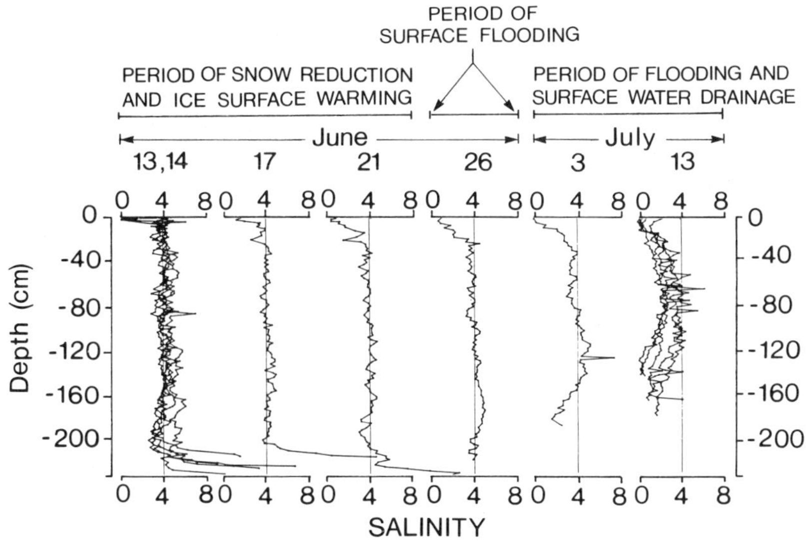

Fig. 3.11 Change in salinity profile for first-year ice during rapid desalination period (Holt *et al.*, 1985). Reproduced by permission of the American Geophysical Union.

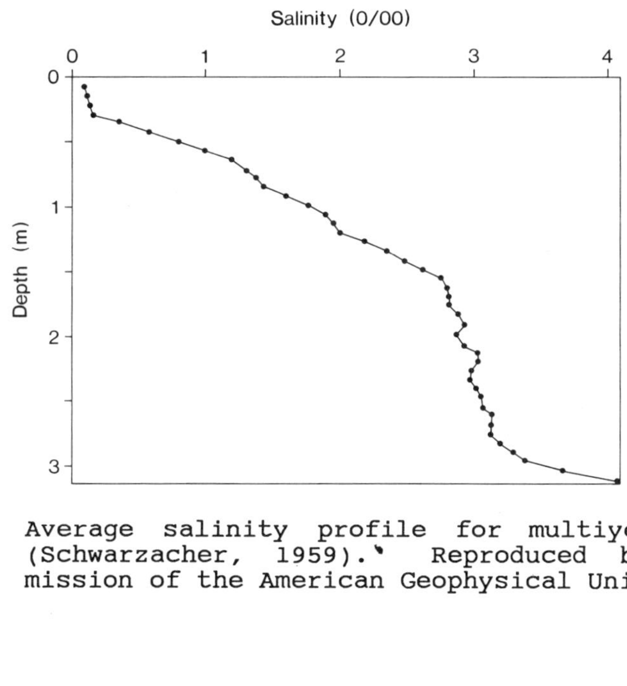

Fig. 3.12 Average salinity profile for multiyear ice (Schwarzacher, 1959). Reproduced by permission of the American Geophysical Union.

measurements agree favourably with more recent results (Campbell *et al*., 1978; Cox and Weeks, 1974). Figure 3.13 shows contours of salinity for a multiyear floe where the salinity of the ice above the mean sea level is generally low. However, the salinity in the ice depressions can remain quite high as shown on the left and right sides of Figure 3.13, whereas the salinity on the hummock in the centre is considerably lower.

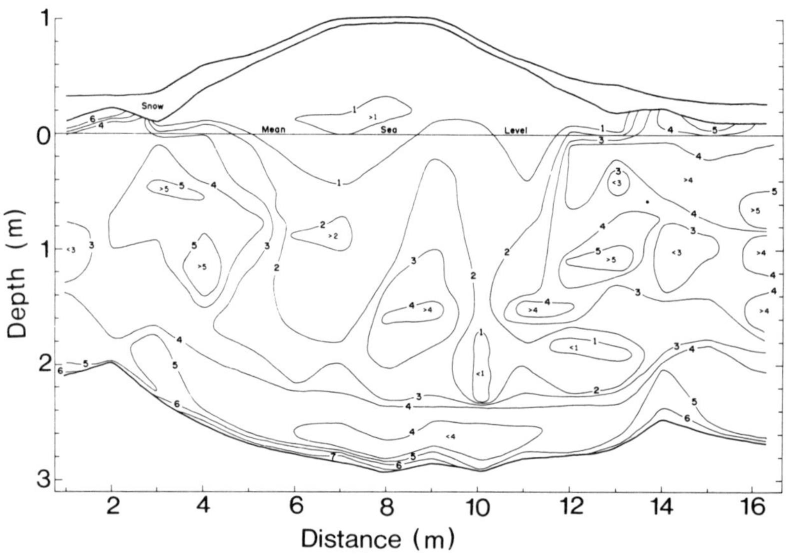

Fig. 3.13 Salinity contours in multiyear ice floe. Contour lines are at $1^{\circ}/_{oo}$ intervals (Cox and Weeks, 1974). Reproduced by courtesy of the International Glaciological Society from *Journal of Glaciology*, *13*(67), p. 109-120.

Temperature

The temperature in the ice increases linearly from the atmospheric temperature at the ice/air interface to about -1.8°C at the ice/water interface. Plot A of Figure 3.14 shows the temperature structure for first-year ice, while

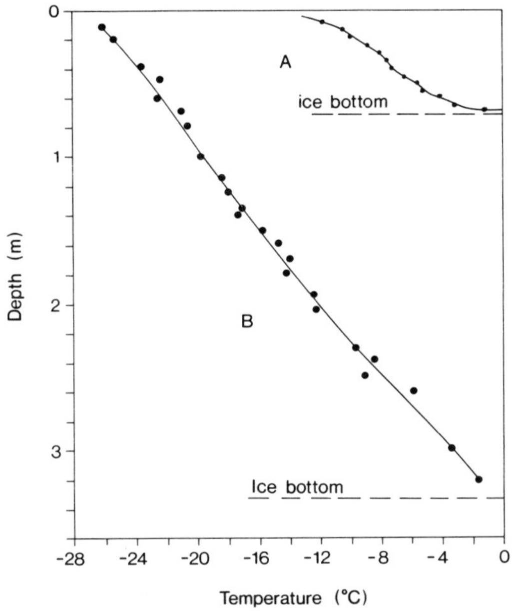

Fig. 3.14 Temperature profile for first-year (A) and multiyear (B) ice. (Plot A from Martin, 1979; Plot B from Cox and Weeks, 1974.) Reproduced by courtesy of the International Glaciological Society from <u>Journal of Glaciology</u>, <u>22</u>(88), p. 473-502 and <u>13</u>(67), p. 109-120.

Plot B shows a similar temperature structure for multi-year ice.

<u>Density</u>

The density of sea ice decreases with age as the brine cells are replaced by ice (Weeks and Lee, 1958). Figure 3.15 shows the density of ice as a function of temperature and salinity. Weeks and Lee (1958) reported densities of newly formed ice at Hopedale, Labrador, as high as 0.945 g/cm^3. Schwartz and Weeks (1977) and Evans (1965) reported densities of first-year ice in the range of 0.910 to 0.920 g/cm^3, while values for multiyear ice

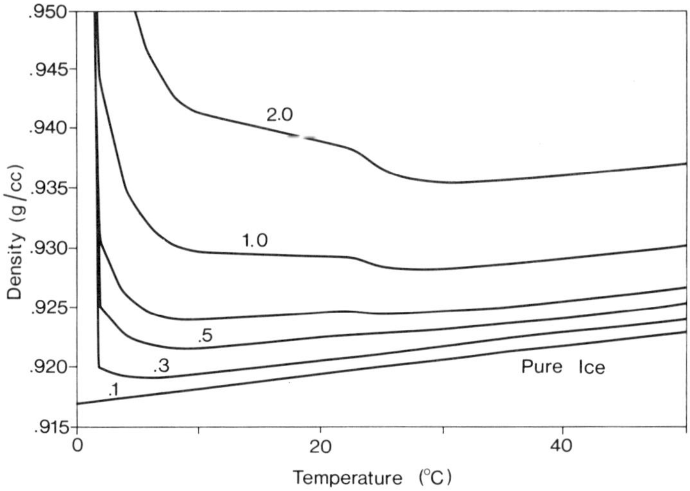

Fig. 3.15 Density of sea ice versus temperature and salinity (data from Anderson, 1960).

are quoted as 0.910 to 0.915 g/cm³. Evans (1965) notes that for multiyear ice, the higher the freeboard, the lower the density, as given by:

$$\rho = -194f + 974 \tag{3.1}$$

where ρ is the density in kg/m³ and f is the freeboard in metres. The freeboard is the height of the ice surface above the mean sea level. These density values compare to a density of seawater of 1.026 g/cm³ at 5°C.

For the top 10 to 20 cm of very porous second-year ice, Sinha[1] has found densities as low as 0.65 g/cm³. Cox <u>et al</u>. (1984) reported average densities of 0.875 g/cm³ in pressure ridge sails and 0.899 g/cm³ in the keels. These low densities, not atypical for the top surface of multiyear ice, are the result of a great number of relatively large air bubbles and voids. Sinha reported that

[1] Sinha, verbal communication, 1985.

near the surface, the bubble diameters are in the range of 4 to 6 mm, while at the 10 cm depth they are about 3 mm. There is a sharp decrease in size to 0.4 mm at 15 cm, with most voids in the lower columnar ice being in the range of tenths of millimetres. Vant et al. (1978) assume spherical bubbles of 5 mm diameter for multiyear ice for modelling purposes.

There are very few reports of bubbles in first-year ice, although Nakawo (1983) reported values of 0.5 to 1 mm as typical. Vant et al. (1978) assumed the value of bubble sizes in first-year ice to be 0.5 mm.

Brine Volume

The volume of brine in the ice determines the salinity of the ice, and varies directly with the temperature of the ice. The brine volume not only affects the electrical properties of the ice, but also significantly affects the structural properties and, in particular, ice strength.

Frankenstein and Garner (1967) have formulated an equation for brine volume in parts per thousand (°/₀₀) for the temperature range from -0.5 to -22.9°C:

$$V = S\left(\frac{49.185}{T} + 0.532\right) \quad (3.2)$$

where S is the salinity of the ice in °/₀₀, and T is the ice temperature in °C. As expected, the brine volume is directly proportional to the salinity of the ice and inversely proportional to the temperature.

Schwartz and Weeks (1977) summarized the work of other investigators to show that compressive, tensile, and flexural strength are all inversely related to the square root of the brine volume. Rossiter (1980) showed a relationship for ice strength as a function of brine volume derived from Schwartz and Weeks (1977) as:

$$\sigma = b_0 \left(1 - \frac{V}{b_1}\right)^{\frac{1}{2}} \tag{3.3}$$

where σ is a measure of ice strength and b_0 and b_1 are constants.

Anderson (1960) reports that "under natural conditions, young sea ice is generally only one-third to one-half as strong as fresh ice but increases in strength as it freshens during subsequent winters."

Electrical Properties

The electrical property of sea ice that is most important in the radar sensing of ice is the relative complex permittivity:

$$\varepsilon_r = \varepsilon_r' - j\varepsilon_r'' \tag{3.4}$$

where ε_r' is the dielectric constant and ε_r'' is the dielectric loss. The loss tangent:

$$\tan \delta = \frac{\varepsilon_r''}{\varepsilon_r'} \tag{3.5}$$

is directly related to the attenuation of the radar signal in the medium. According to Morey *et al.* (1984), there are two mechanisms that cause dissipation of the radar energy as heat, namely, relaxation and conduction. Conduction, which causes losses at low frequencies, is caused by electron and ion movement in an electric field. At higher frequencies, above about 1 GHz, relaxation losses caused by the drag of oscillating dipolar water molecules constitute the major losses. Relaxation losses are totally dependent on free water content.

The value of ε_r' for fresh ice has been measured extensively: 3.17 at 10.0 GHz (Lamb, 1949), 3.15 at 9.375 GHz (Cumming, 1952), and 3.17 at 10.0 GHz (Von Hippel, 1954).

The value of ε_r' for saline first-year ice is higher, and is highly dependent on the temperature and salinity of the ice. First-year sea ice is a complex mixture of pure ice and brine. Although there is only a small quantity of brine present in the ice, its large dielectric constant (ten to twenty times that of the ice itself) has a significant influence on the resulting dielectric properties of the ice-brine mixture. Figure 3.16 shows the apparent dielectric constant for first-year ice as a function of brine volume and depth. Kovacs and Morey (1979) reported a value of 4.6 for ε_r' for first-year ice.

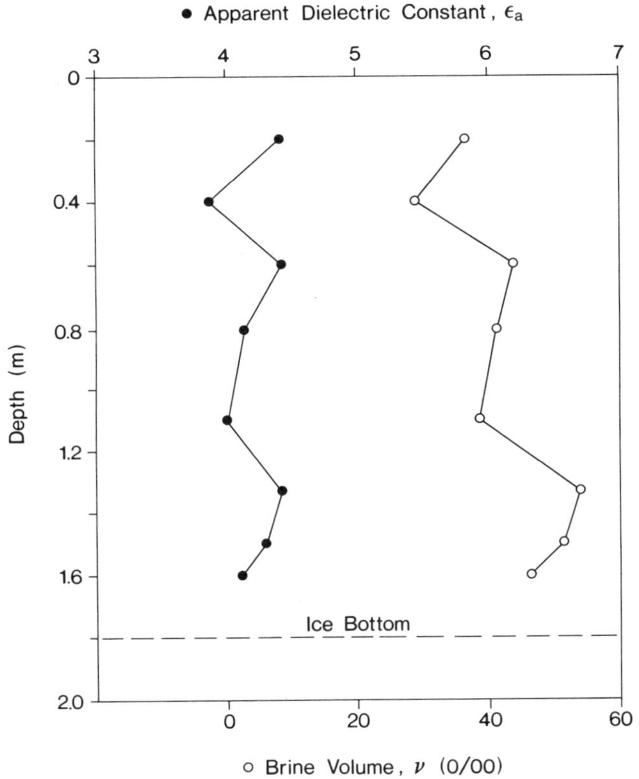

Fig. 3.16 Apparent dielectric constant for first-year ice versus brine volume and depth (Morey <u>et al</u>., 1984).

Figure 3.17 shows the relationship between ε_r' and temperature. The value of ε_r' for multiyear ice is lower than that for first-year ice and is much less temperature sensitive. Kovacs and Morey (1978, 1979) reported a value of 3.7 for multiyear ice.

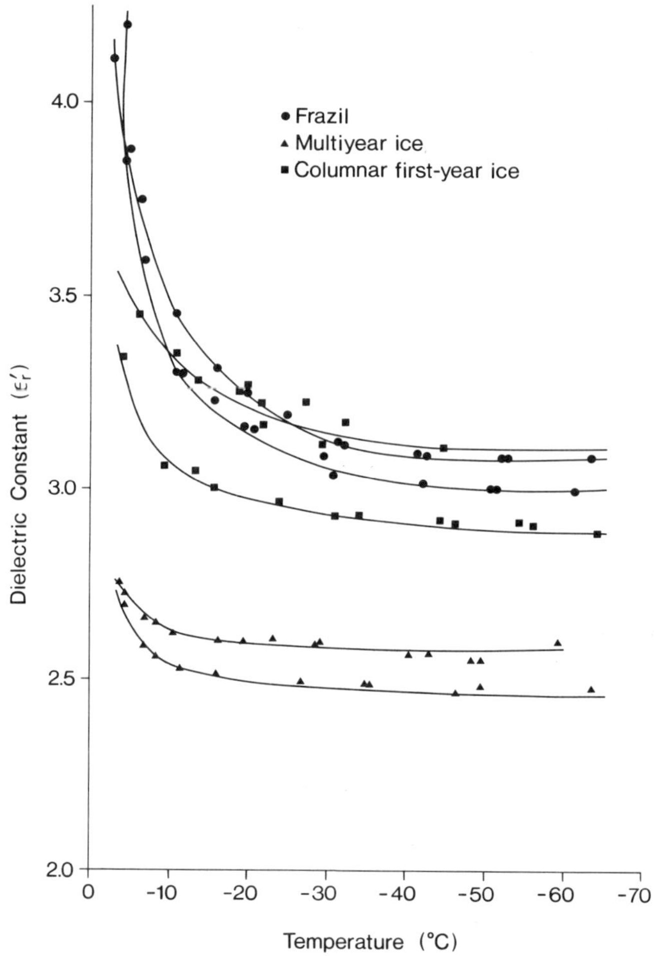

Fig. 3.17 Dielectric constant versus temperature for sea ice (Vant et al., 1974). Reprinted by permission of the American Institute of Physics.

The loss tangent for pure ice has been measured by several investigators: 20 x 10^{-4} (Vant *et al.*, 1974), 12 x 10^{-4} (Lamb, 1949), 27 x 10^{-4} (Cumming, 1952), and 7 x 10^{-4} (Von Hippel, 1954), all at slightly different temperatures.

Unlike the dielectric constant, which is relatively independent of frequency above about 10^6 Hz (1 MHz), the dielectric loss for sea ice is strongly dependent on frequency. This condition results from the fact that both liquid brine and water inclusions exist in sea ice, and because the dielectric properties of water are strongly dependent on frequency in the range of 10^9 to 10^{11} Hz (Hoekstra and Spanogle, 1972). Figure 3.18 shows this clearly. In the range of 3 to 10 GHz (X- and S-band marine radar) the dielectric loss is at its minimum.

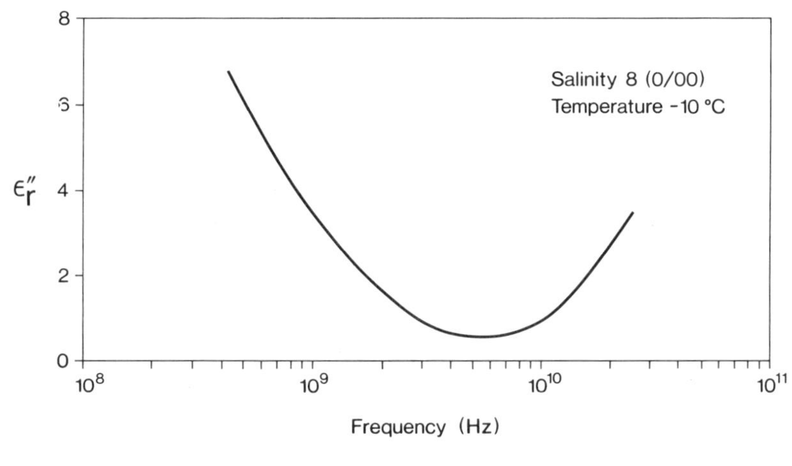

Fig. 3.18 Dielectric loss versus frequency for first-year ice (Hoekstra and Spanogle, 1972). Reprinted by permission of the Cold Regions Research and Engineering Laboratory.

As in the case of ε_r', the dielectric loss for sea ice is also strongly temperature-dependent. Figure 3.19

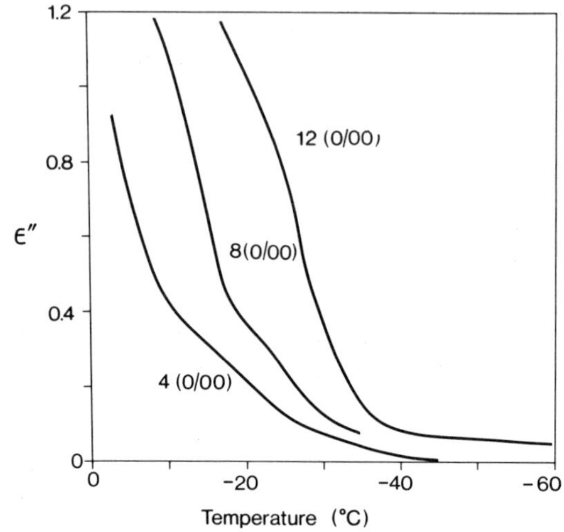

Fig. 3.19 Dielectric loss versus temperature for several salinities of sea ice (Hoekstra and Spanogle, 1972). Reprinted by permission of the Cold Regions Research and Engineering Laboratory.

shows the dielectric loss versus temperature for several salinities. The strong increase in dielectric loss as the temperature increases is a result of the precipitated salts going back into solution. In the normal range of ice temperatures encountered (-30 to 0°C), the dielectric loss increases substantially and is very temperature sensitive.

The dielectric loss is much lower for multiyear ice, because of its lower salinity, and much less temperature sensitive. Figure 3.20 shows the variation of ε_r'' with temperature for multiyear ice.

As in the case of the strength of ice, the dielectric properties can be shown to be related to the brine volume. Hoekstra and Cappillino (1971) have shown that, at 9.8 GHz, ε_r' can be determined by:

$$\varepsilon_r' = \frac{(\varepsilon_r')_{ice}}{1 - 3\,V_{br}} \tag{3.6}$$

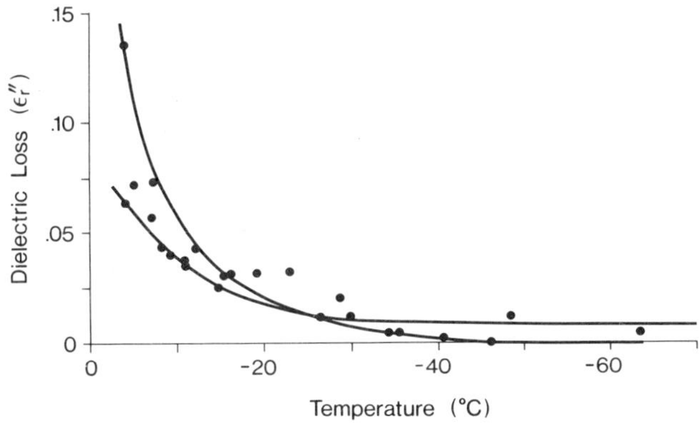

Fig. 3.20 Dielectric loss versus temperature for multiyear ice. Salinity 0.61 to 0.70 ppt (Vant et al., 1974). Reprinted by permission of the American Institute of Physics.

where $(\varepsilon_r')_{ice}$ is the dielectric constant for pure ice, and V_{br} is the percent brine volume. Vant et al. (1978) have also developed the relationship (at 10 GHz):

$$\varepsilon_r' = a_0 + \frac{a_1}{1 - 3 V_{br}} \qquad (3.7)$$

where the best fit to the data for columnar first-year ice, frazil first-year ice, and multiyear ice showed that $a_0 \neq 0$ and $a_1 \neq 3.14$. This result was interpreted to mean that all the brine was not in spherical inclusions, but that perhaps the brine was held in elliptical inclusions angled to the electrical field. Vant et al. (1978) also showed that ε_r'' is directly related to V_{br} and went on to examine other empirical and theoretical models.

Through the use of a theoretical model and correlation with real data, Vant et al. showed that ε_r' and ε_r'' varied with the angle of inclination of the assumed elliptical brine pockets. Figure 3.21 plots the theoretical model and real data, and shows the best agreement when the brine pockets are oriented at 35 to 45° to the vertical.

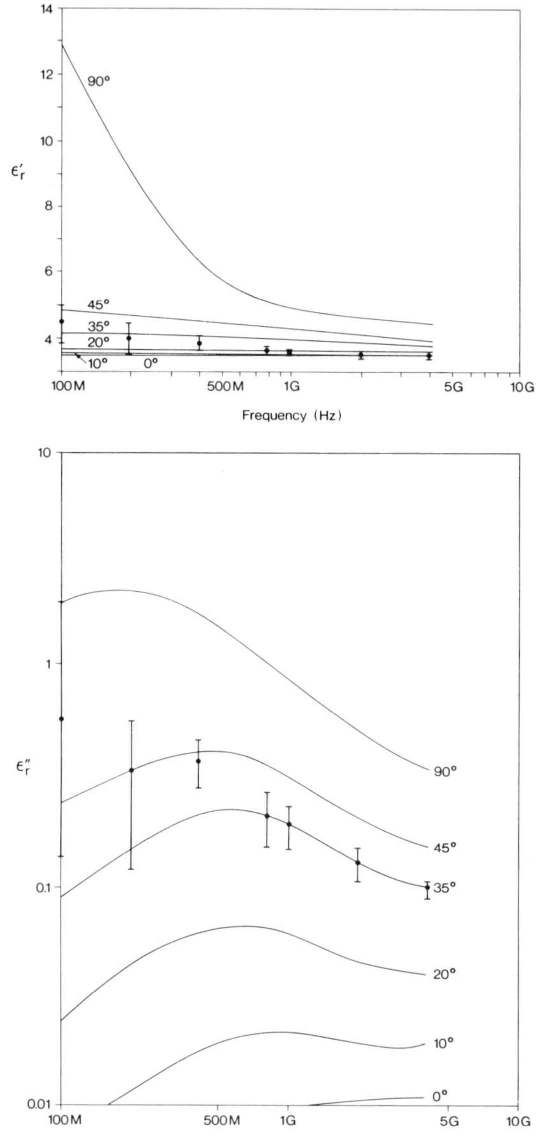

Fig. 3.21 Comparison of theoretical model (solid line) with experimental data for ε'_r and ε''_r (Vant et al., 1978). Reprinted by permission of the American Institute of Physics.

Table 3.1 Skin depth at X-band as a function of loss tangent.

Loss tangent ($\tan\delta$)	Ice type	Skin depth (cm)
0.001	Glacier ice (0.001–0.005)	541.0
0.005		108.0
0.01	Multiyear ice (0.01–0.06)	54.1
0.02		27.0
0.04	First-year ice (0.04–0.30)	13.5
0.06		9.0
0.08		6.8
0.10		5.4
0.15		3.6
0.20		2.7
0.30		1.8
0.40		1.4
0.50		1.1

The loss tangent dictates how quickly the transmitted signal dies out as it propagates through the ice. From Evans (1965), the field strength of the electromagnetic field decreases with exponential α, where:

$$\alpha = \frac{2\pi}{\lambda} \left\{\frac{\varepsilon_r'}{2} \left(\sqrt{1 + \tan^2\delta} - 1\right)\right\}^{1/2} \qquad (3.8)$$

The skin depth, $1/\alpha$ (where the incident electromagnetic energy has been reduced to $1/e$) defines the depth of penetration of an electromagnetic wave into ice. Table 3.1 shows the skin depth as a function of loss tangent at X-band (3-cm wavelength). The attenuation of the signal is measured in dB/m; it is given by Vant *et al.* (1978) as:

$$\text{Attenuation} = 868.6\alpha \tag{3.9}$$

From this equation, and from Table 3.1, we can see that the higher losses in first-year ice significantly reduce the penetration into the ice, while the lower losses in multiyear ice will allow significant penetration. This condition was noted by Weeks and Ackley, (1982) who defined young and first-year ice as a "high loss" and multiyear ice as a "low-loss" dielectric material.

<u>Reflection Coefficient</u>

As stated by Page and Ramseier (1975), "radar reflections occur when there is an abrupt change in the value of ε_r, i.e., where significant changes in ε_r take place within the distance of a wavelength of the radio wave." Budinger <u>et</u> <u>al</u>. (1960) related the reflection coefficient, ρ, to the dielectric constant as:

$$\rho = \frac{\sqrt{\varepsilon_r'} - 1}{\sqrt{\varepsilon_r'} + 1} \tag{3.10}$$

We can, therefore, expect reflections from air/ice interfaces, ice/water interfaces (if the losses in the ice are low enough to permit penetration), ice/air interfaces (for low losses) from the back surface of the ice, and from inhomogeneities in the ice.

3.2 GLACIER ICE

Glacier ice, unlike sea ice, has its origin on land. The glaciers spawn billions of tons of icebergs each year from tongues which reach the sea. The icebergs drift over an area of 70,000,000 km², or more than 20% of the ocean area, and pose a serious threat to navigation and offshore activity in many areas of the world. The "unsinkable" ocean liner <u>Titanic</u> collided with an iceberg, spawned in West Greenland, in the north Atlantic in April 1912 with the loss of 1,513 lives. The Titanic did not carry radar. More recently, the Danish ice-breaking passenger cargo ship Hans Hedtoft sank off Greenland in 1959 with a loss of 95 lives after collision with an iceberg. The Hans Hedtoft did carry radar.

Morphology and Density

Glaciers form where precipitation in the form of snow exceeds the evaporation. The formation of glacier ice starts with snowfall on the high altitude areas of existing glaciers. The new-fallen snow has a low density (0.06 to 0.08 g/cm³) because of air trapped between the hexagonal snow crystals.

The length of time for the snow to be converted into ice depends mostly on temperature. In temperate glaciers (glaciers where the ice is at the pressure-melting point throughout) the delicate points of the snow crystals begin to melt quickly, the snowflakes are converted to spherical particles, the snow settles, and the density increases. The density can increase to 0.2 g/cm³ in a few days. This more dense snow is granular in form. However, in sub-polar and polar glaciers, this melting and compression process may take years. The compaction process is accompanied by, and is a result of, the ice crystals growing larger, joining together, and thus eliminating the air spaces between the crystals. As the snow continues to become more dense through compression, recrystallization, and freezing and thawing, it reaches an intermediate step en route to becoming glacier ice. This intermediate step, known as firn or névé, occurs when the snow reaches a density of between 0.4 and 0.55 g/cm³.

As the density increases further, mostly by recrystallization, and the interconnecting air passages between ice grains are sealed off, the firn is transformed into glacier ice. The density at which this occurs is around 0.8 to 0.85 g/cm³. For Greenland glaciers this process may take from 150 to 200 years after the original snowfall. As the firn is transformed into ice, the air that was trapped in the firn is either forced out through cracks or fissures in the ice or is trapped in bubbles. The further increase in density of the ice results from the compression of these air bubbles. The pressure in air bubbles within Greenland glacier ice has been measured as high as 10 to 15 atmospheres (Scholander and Nutt, 1960). The glacier ice is impermeable to air or water.

The visible "whiteness" of glacier ice is caused by tiny air bubbles distributed throughout the ice. These bubbles give glacier ice a 4 to 9% covolume of air. Scholander and Nutt (1960) reported that for Greenland icebergs, in general, the bubbles are elongated with diameters of 0.02 to 0.18 mm, and lengths of up to 4 mm. They also reported a lesser number of spherical bubbles with diameters ranging from microscopic up to 2 mm. Gow (1968), in a detailed analysis of bubble content in Antarctic glaciers, reported a range of spherical bubble sizes of 0.49 to 0.33 mm, decreasing with depth. The density of the bubbles, however, was relatively independent of depth. Gammon _et al_. (1983) reported round or ellipsoidal bubbles, with elongation in the order of a few tenths of millimetres. It would seem that typical values for bubble sizes in glacier ice would be in the order of 0.1 to 0.5 mm. The occasional piece of bluish glacier ice indicates ice which is free of bubbles.

The final density of glacier ice can exceed about 0.9 g/cm³. As seawater has a specific gravity (or density) of approximately 1.0 g/cm³, about six-sevenths of the mass of floating glacial ice will be below the sea surface, while the freeboard of individual icebergs will depend on their shape.

Electrical Properties

The dielectric constant for glacier ice is about 3 at 10 GHz (Kirby and Lowry, 1979). Measurements by Pearce and Walker (1967) at 30 MHz for an ice density of 0.92 g/cm^3 yielded a dielectric constant of 3.31 ± 0.04, while Jezek *et al*. (1978) reported values between 3.09 and 2.89. These values, coupled with the fact that glacier ice is not saline, make it a low-loss dielectric. Page and Ramseier (1975) reported that the loss tangent had values less than 2×10^{-3} for all frequencies greater than 1 GHz.

Icebergs and Their Fragments

Under the pressure of its own weight, the glacier flows from its high altitude source towards sea level. Glacier flow in Greenland has been measured at up to 20 m/day. In many instances the glacier continues to flow out over the sea, often reaching distances of 40 to 60 km off shore. Under the continuing action of the tides and ocean level changes caused by winds and swell, large pieces of ice break off the glacier tongues.

Depending on their size, these ice features are classified as either ice islands or icebergs. Ice islands are large pieces of ice which usually exceed 5 m above sea level, are 30 to 50 m thick, and are from a few thousand square metres to 500 km^2 or more in area. Ice islands are relatively rare in the Arctic. Icebergs are pieces of ice higher than 5 m above water level, but covering less area than an ice island.

Each year, west Greenland, and, to a lesser extent, eastern Canadian arctic glaciers, produce about 10,000 icebergs. The icebergs are trapped by sea ice in the arctic fjords until they are released during breakup. The icebergs then begin a long and circuitous route which can take from two to five years, until their final disintegration, possibly somewhere off Newfoundland.

As the icebergs drift and begin to deteriorate, they fracture and break, or calve off smaller pieces. Pieces which are from 1 to 5 m above sea level are referred to as bergy bits, while smaller pieces, up to 1 m above sea level, are called growlers. Table 3.2 lists the representative height above water level, waterline length, and physical area for icebergs, bergy bits, and growlers. Appendix A gives more complete definitions for icebergs and other ice types.

The icebergs are swept along by the counter-clockwise Baffin Bay circulation into northern Baffin Bay, and then they flow southward off Baffin Island, driven by the relatively narrow but rapidly flowing Baffin current. Icebergs tracked by satellite (Marko et al., 1982) have

Table 3.2 Classification of icebergs by size.

Type	Height above water-line (m)	Water-line length (m)	Physical area above water-line (m^2)	Relative size	Mass (tonnes)
Iceberg	>5	>30	>300	merchant ship	>180,000
Bergy bit	1-5	10-30	100-300	small house	up to 5,400
Growler	<1	<10	<100	grand piano	up to 120

Icebergs are further classified by size as follows:

Small:	Height less than 16 m, length less than 65 m.
Medium:	Height 16 to 48 m, length 65 to 130 m.
Large:	Height 48 to 70 m, length 130 to 225 m.
Very large:	Height greater than 70 m, length greater than 225 m.

shown long-term drift rates of up to 12 km per day. However, the icebergs routinely make excursions into Lancaster Sound, on occasion venture into Navy Board Inlet, regularly become trapped in landfast ice, or run aground. Although Marko et al. (1982) tracked four icebergs which travelled from Lancaster Sound to Davis Strait in only 8 to 15 months, they suggested that the most likely duration for an iceberg to drift from Lancaster Sound to northern Labrador (66°N) would be in excess of three years.

The incidence of icebergs decreases farther south: Figure 3.22 shows that there is an almost linear decrease in the number of icebergs from northern to more southern latitudes. The number of icebergs that cross 48°N annually and enter the Grand Banks area is shown in Figure 3.23 (data from the International Ice Patrol).

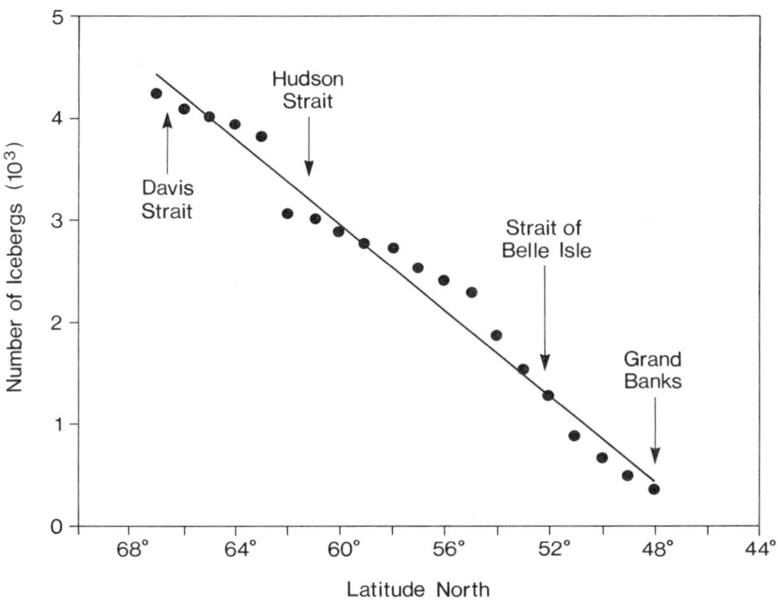

Fig. 3.22 Annual average number of icebergs (dots) crossing selected latitudes (Ebbesmeyer et al., 1980).

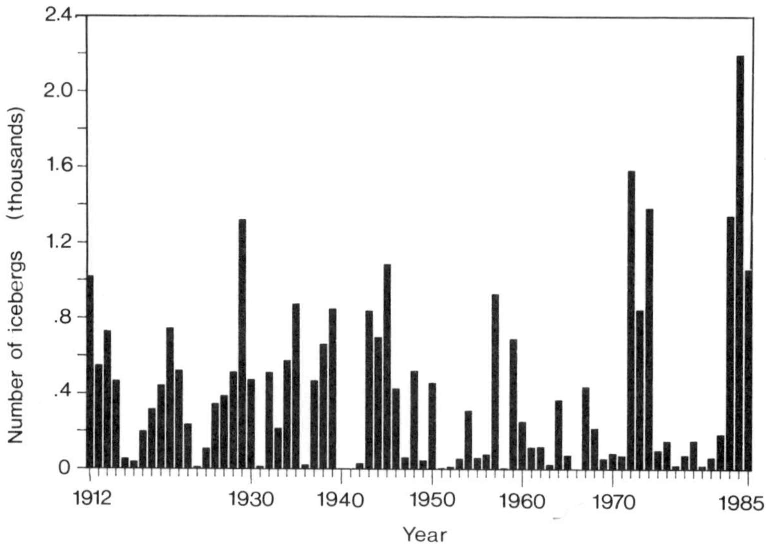

Fig. 3.23 Number of icebergs crossing 48°N (data from International Ice Patrol).

The number has been as high as over 2,000 in 1984 and as low as none in 1966. The relatively large number of icebergs reported in recent years, however, is partially the result of better surveillance and detection techniques.

The deterioration of icebergs is slow in Baffin Bay and along the Labrador coast because of the near-freezing seawater temperatures. However, as the icebergs move farther south into warmer water they deteriorate more rapidly. When the icebergs reach the vicinity of 48°N (northern Newfoundland) they have lost more than 85% of their mass (Davidson and Denner, 1982). Few icebergs pass south of 44°N and icebergs are rarely reported south of 40°N.

3.3 SUMMARY

Those characteristics of the ice types of interest (first-year, multiyear, glacier) which may affect their remote sensing by surface-based radars have been identified.

First-year sea ice is relatively saline (5 to 20‰) and thus is lossy and difficult to penetrate with microwave energy. This ice is usually columnar in structure, with a possible thin layer of frazil ice on the surface. The surface topography is generally sharp and angular with moderate relief.

Multiyear ice is a low-loss dielectric substance which permits significant penetration by microwave energy. The top layer (about 1 m) of the ice is porous with relatively large (2 mm) air voids. The surface is undulating with smooth and rounded features, often with significant vertical relief.

Glacier ice is also a low-loss dielectric material which permits significant microwave penetration. It usually contains a large number of small (0.2 to 0.5 mm) air bubbles. Surface features of glacier ice on the ocean (icebergs, bergy bits, and growlers) vary from angular and square to smooth and well rounded.

CHAPTER 4
Radar Theory

The application of radar to surveillance of the marine environment is reasonably well understood, although the extension of this knowledge to ice applications is not well documented. This chapter will review the salient radar considerations as they apply to the surface-based sensing of ice.

4.1 FREE-SPACE RADAR EQUATION

An appropriate starting point to discuss radar theory is the radar equation, the free-space form of which is (Skolnik, 1980):

$$P_r = \frac{P_t \, G^2 \, \lambda^2 \, \sigma_t \, F^4}{(4\pi)^3 \, R^4 \, L} \qquad (4.1)$$

where P_r = power received
P_t = power transmitted
G = antenna gain
λ = radar wavelength
F = pattern propagation factor (equals 1 for free space)
L = system losses
σ_t = radar cross-section of target
R = range to target.

In the strictest sense, free space means that the region illuminated by the transmitted signal is void except for the radar and the target. In a practical sense, free space propagation can be considered to occur in a clear atmosphere in the absence of turbulence, well removed from the earth's surface. The effect of the pattern propagation factor (F) will be discussed in Chapter 6. From Equation 4.1 it can be seen that for the free space case, the power received is inversely proportional to the fourth power of range, directly proportional to the square of the wavelength of the transmitted signal, and directly proportional to the radar cross-section of the target.

4.2 RADAR CROSS-SECTION

The value of σ_t from Equation 4.1 is a measure of the portion of the incident energy reflected back to the radar. By definition, the radar cross-section is the strength of a radar target expressed in terms of the cross-section of an ideal reflector which would give the same returned signal strength. The radar cross-section has units of area. The signals returned from a target vary with wavelength of transmitted signal, target geometry, aspect, and reflectivity, such that the computed radar cross-section for a given target can vary significantly.

As an example of the effect of geometry on the radar cross-section of a target, consider the case of a sphere and a flat circular plate, both having a radius of 1 m. The targets are perfect conductors and the wavelength (3 cm) is small compared to the dimensions of the targets.

For the sphere, the radar cross-section is defined in the optical region of the plot in Figure 4.1 (Skolnik, 1970), since $2\pi r/\lambda$ is greater than 200. Therefore, its

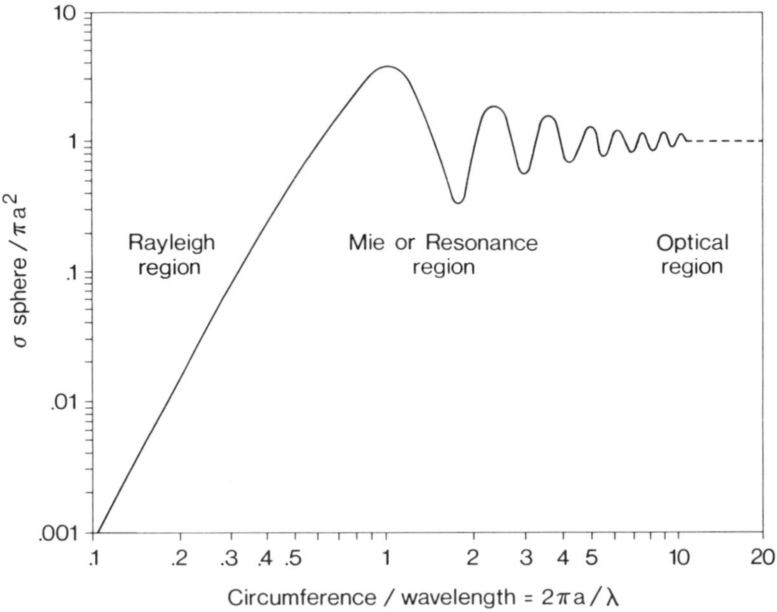

Fig. 4.1 Variation of radar cross-section of a sphere as a function of wavelength or frequency from Skolnik (1980), Introduction to Radar Systems, McGraw-Hill.

radar cross-section is its projected optical cross-section of πr^2 m^2, and for the sphere of 1 m radius this is πm^2. As the sphere remains the same when viewed from any direction, the radar cross-section is not sensitive to aspect.

Conversely, when the flat plate is normal to the radar beam, the radar cross-section can be shown to be (Blake, 1980):

$$\sigma = \frac{4\pi A^2}{\lambda^2} \tag{4.2}$$

where A is the optical area in square metres. For a circular plate of area πm^2 and a wavelength of 3 cm, the radar cross-section becomes 138,000 m^2. This cross-section, however, is realized only when the plate and the radar beam are perpendicular. Williams (1979) states that as the plate is inclined away from the beam, the reflected pattern takes the form of (sin x)/x, such that the first side-lobe peak is only 13 dB down (6,000 m^2) and in practice (because of blurring) the first null will be no more than 30 dB down (138 m^2).

In the application considered here, ice targets with flat faces (new pressure ridges, rafted faces, and icebergs with flat surfaces) will appear to have substantially larger radar cross-sections than ice features with spherical or rounded surfaces (weathered pressure ridges, hummocked ice, etc.) even if the flat surfaces are at an angle to the radar beam. However, it is very difficult to predict the radar cross-section of ice targets because of their random shape, height, and surface roughness.

4.3 INCIDENCE AND GRAZING ANGLES

In describing the geometry of the interaction of a radar signal and a surface target, three terms are used: incidence angle, depression angle, and grazing angle. These are shown in Figure 4.2. The depression angle is defined as the angle of radiation with respect to the local horizon at the transmitter. The incidence angle is defined as the angle between the impinging radar beam and a perpendicular to the surface at the target. The grazing angle is defined as the angle with respect to the local horizon at the target and is the complement of the incidence angle.

For normal propagation conditions, radio waves are refracted downward towards the earth's surface, resulting in slightly longer ranges of propagation than would normally be expected with a straight-line path. To account for this extended range, the earth's radius has been

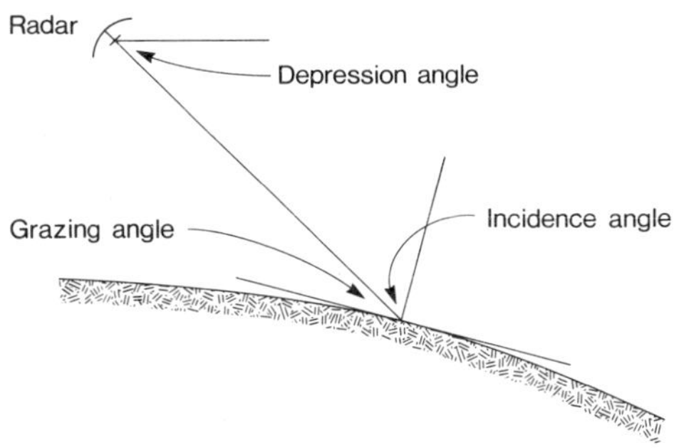

Fig. 4.2 Angles used to describe surface-based radar operation.

extended by a 4/3 factor to yield an effective earth's radius, which accounts for "standard" propagation conditions.

Given this 4/3 earth approximation for the radar propagation path, Blake (1980) has derived expressions for the grazing angle, ψ, and the depression angle, α, as follows:

$$\psi \simeq \sin^{-1}\left(\frac{h}{R} - \frac{R}{2a_e}\right) \qquad (4.3)$$

and

$$\alpha = \sin^{-1}\left(\frac{h}{R} + \frac{R}{2a_e}\right) \qquad (4.4)$$

where h = height of radar antenna
R = range
a_e = 4/3 earth radius.

The sines of the two angles, ψ and α, differ only by the factor R/a_e. For ranges up to 10 nm, the two angles differ by less than 2/10 of a degree. Therefore, for surface-based radar, these two angles can be considered to be effectively the same. The term grazing angle will be used throughout this work. Figure 4.3 shows a plot of grazing angle versus range for typical antenna heights for surface-based radar work.

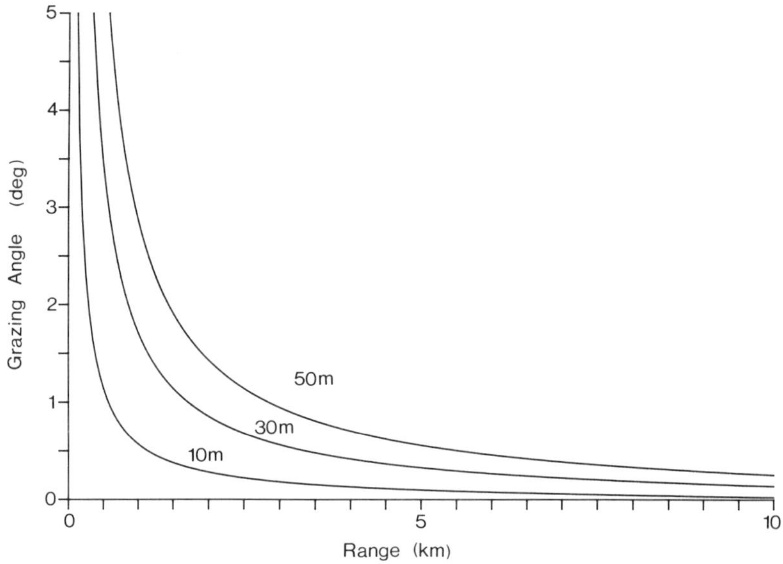

Fig. 4.3 Grazing angle versus range for surface-based radar. Antenna heights of 10, 30, and 50 m.

For surface-based radar, an even more important definition is the term "local" grazing angle, which accounts for the effect of the local surface topography. As indicated earlier, the grazing angle is the angle the radar wave makes with respect to the horizon at the target. As shown in Figure 4.4, for a smooth surface the grazing angle for a surface-based radar is small. For a rough surface the grazing angle is also small; however, the "local" grazing angle between the tangent to the surface at the target and the radar wave can be considerably larger. This larger "local" grazing angle results in significantly more energy being returned to the radar.

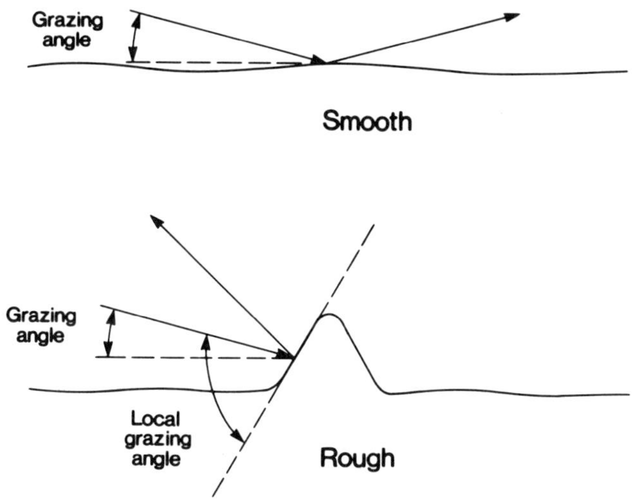

Fig. 4.4 Designation of "local" grazing angle.

4.4 AREA CLUTTER RADAR EQUATION

In the case where the radar illuminates the surface of the earth or the ocean, not only is the target illuminated, but also the surface itself. In this case, the power returned is from both the target of interest and from the surface.

In the situation where the target is large compared to the radar resolution cell, or is made up of a large number of individual scatterers, σ_t in Equation 4.1 must be modified to take into account the distributed nature of the target. Signal returns from a distributed target of this nature are often referred to as clutter, since they mask point target echoes. In some instances the clutter is actually the target of interest, as in the classification of ice features. In other instances (i.e., the detection of a growler in the sea), the clutter from the sea is an impediment to detection. Hence, the role of clutter varies with the application of interest.

The surface area illuminated, referred to as the resolution cell, is equal to the lateral extent of the radar illumination times the range extent. The lateral extent of the area illuminated is determined by the horizontal beamwidth of the antenna, θ_B, while the range extent is determined by the pulse length, τ, and the grazing angle, as shown in Figure 4.5. The area illuminated then becomes

$$A = \frac{c\tau}{2} R \theta_B \sec \psi \qquad (4.5)$$

where c = speed of light
τ = pulse length
R = range (metres)
θ_B = horizontal beamwidth (radians)
ψ = grazing angle.

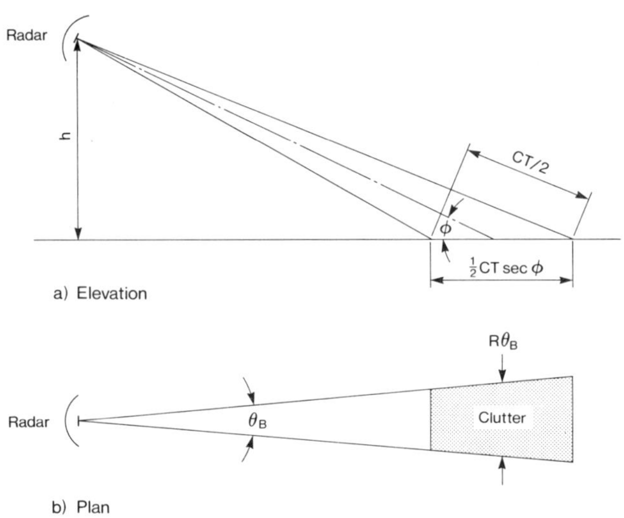

Fig. 4.5 Geometry of radar clutter cell.

The area illuminated (resolution cell) as a function of range is shown in Table 4.1.

Table 4.1 Grazing angle and resolution cell size versus range for a typical marine radar.

Range (nm)	Grazing angle (°)	Resolution cell* (m^2)
0.5	2.3	4605
1.0	1.1	9210
2.0	0.57	18420
3.0	0.38	27630
4.0	0.28	36841
5.0	0.23	46051
6.0	0.19	55261

* Resolution cell for a 1 μsec pulse with 1.9° horizontal beamwidth at an antenna height of 35 m.

The clutter cross-section, σ_C, is equal to the areal extent of the radar illumination (Equation 4.5) times the normalized radar cross-section, σ_o, of the illuminated area:

$$\sigma_C = A\sigma_o \qquad (4.6)$$

where the normalized radar cross-section is usually with reference to one square metre.

For the low grazing angles associated with surface-based radars, sec ψ in Equation 4.5 is approximately equal to 1, such that the equation for σ_C reduces to:

$$\sigma_C = \frac{c\tau}{2} R\,\theta_B\,\sigma_o \qquad (4.7)$$

Using the expression for σ_C in place of σ_t, Equation 4.1 becomes:

$$P_r = \frac{P_t\, G^2\, \lambda^2\, (c\tau/2)\, \theta_B F^4}{(4\pi)^3\, R^3\, L}\, \sigma_0 \qquad (4.8)$$

Rearranging this equation to solve for σ_0 yields:

$$\sigma_0 = \frac{P_r\, (4\pi)^3\, R^3\, L}{P_t\, G^2\, \lambda^2\, (c\tau/2)\, \theta_B F^4} \qquad (4.9)$$

Comparison of Equations 4.1 and 4.8 shows that the typical R^4 relationship between σ_t and range (as shown in Equation 4.1 for a point target) has been reduced to an R^3 relationship for a distributed target.

4.5 DETECTION RANGE

For a "normally" refractive atmosphere (one in which the refractive-index gradient decreases linearly with altitude) the radar horizon is defined by:

$$R = 4.11(\sqrt{h_a} + \sqrt{h_t}) \qquad (4.10)$$

where h_a = height of antenna (m)
h_t = height of target (m)
R = range (km).

Table 4.2 shows the radar horizon (in km) for various antenna heights to the surface, and to target heights of 1, 5, and 10 m.

The minimum separation required between two targets for the radar to resolve them individually is termed the radar resolution. In range, the returned signals from the two targets must be separated in time by at least the pulse length (τ in seconds) of the transmitted signal, that is,

$$\text{Range resolution (temporal)} = \tau \qquad (4.11)$$

Table 4.2 Range to radar horizon for targets of different height with varying antenna height.

Antenna height (m)	Range to radar horizon (km)			
	10-m high target	5-m high target	1-m high target	0-m high (surface)
5	22.2	18.4	13.3	9.2
10	26.0	22.2	17.1	13.0
20	31.4	27.6	22.5	18.4
50	42.1	38.3	33.2	29.1
100	54.1	50.3	45.2	41.1
200	71.1	67.3	62.2	58.1

In terms of the physical separation in range that two targets must have, this resolution is expressed as:

$$\text{Range resolution (spatial)} = \frac{c\tau}{2} \qquad (4.12)$$

where c = speed of light. The range resolution is independent of the range of the target from the radar and is influenced only by the length of the transmitted pulse. For a 1 μsec pulse the spatial range resolution is 150 m, while for a 250 nsec pulse the resolution is 37.5 m.

As in the case of range resolution, the azimuthal resolution can be expressed in two ways. The angular azimuthal resolution is uniquely specified by the horizontal beamwidth (θ) of the radar antenna. Thus

$$\text{Azimuthal resolution (angular)} = \theta_B \qquad (4.13)$$

The actual spatial azimuthal resolution (the distance by which two targets must be separated in bearing to be individually resolvable) is not only a function of the horizontal beamwidth of the antenna, but also is directly proportional to the range to the target:

$$\text{Azimuthal resolution (spatial)} = \theta_B R \qquad (4.14)$$

where R is the range to the target, and θ_B is in radians. Hence the spatial azimuthal resolution degrades as the range to the target increases. Table 4.3 shows the azimuthal resolution for several horizontal beamwidths and several ranges.

4.6 TARGET DETECTION

The ability of a radar system to detect a target of interest is influenced by the ratio of the average power of desired signal to the average power of interfering signal received. The interfering signals take one or both of two forms: noise, and clutter. Noise is composed of thermal noise associated with the receiver, as well as many sources of man-made noise. Clutter, in general, is the unwanted signal return from the area illuminated by the radar signal.

Table 4.3 Azimuthal resolution at different ranges and for antennas of varying beamwidths.

Range (nm)	Azimuthal resolution (m) with varying antenna beamwidth			
	1°	1.5°	2°	2.5°
0.5	16	24	32	40
1.0	32	48	64	81
2.0	65	97	129	162
5.0	162	243	323	405
10.0	324	485	647	810
20.0	648	970	1294	1620

Noise-Limited Case

In the noise-limited case, the objective is to choose the radar parameters such that sufficient signal is received to exceed system noise. The only ice-related radar situations which are noise-limited are the detection of small pieces of ice in low sea-states, and the detection of icebergs at far range; in the latter case the problem of detection may be compounded by the curvature of the earth and the inability to illuminate the target.

In the noise-limited case, detection is limited by receiver noise where the receiver noise power is given by:

$$P_n = F_n kTB \qquad (4.15)$$

where k = Boltzmann's Constant
= 1.38×10^{-23} W/(Hz°K)
T = receiver temperature (°K)
B = receiver bandwidth (Hz)
F_n = receiver noise figure.

The term kTB is the thermal noise power generated at the receiver input. For a typical marine radar receiver with bandwidth of 10 MHz, the thermal noise power is about -104 dBm. Thus for a practical receiver with a noise figure of 10 dB, the minimum detectable signal level is -94 dBm. This power level is termed the minimum detectable signal (MDS) of the radar, and indicates the target power required for detection. The power returned from a target and from clutter was given in Equations 4.1 and 4.8, respectively. The expressions are for the mean power received. However, target detection performance also depends on the statistical nature of the target, clutter, and noise returns.

These signals are statistically characterized by probability density functions (pdf). The distribution of noise at the front end of a radar receiver is typically Gaussian. However, after rectification by a video detector the noise amplitude becomes Rayleigh distributed. Figure 4.6 shows plots of typical pdfs for noise and signal-plus-noise. Target detection is accomplished by setting a threshold between the means of the pdfs, and a target is declared whenever the threshold is exceeded.

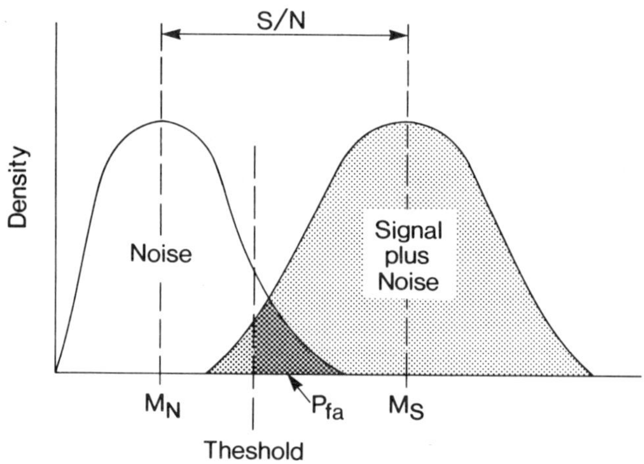

Fig. 4.6 Relationship between P_{fa}, P_d, and S/N ratio.

The integral of the area under the signal-plus-noise curve above the threshold is defined as the probability of detection (P_d), while the integral of the area under the noise curve above the threshold is termed the probability of false alarm (P_{fa}). All noise returns above the threshold will incorrectly be declared as targets. Likewise, the integral of the area under the signal-plus-noise curve below the threshold presents the probability of missed targets as a target below the threshold level will not be declared. Raising the threshold reduces the P_{fa}, but decreases the P_d, increasing the probability of missing a target. In a radar detection system, the threshold is usually chosen so as to provide a constant P_{fa}. This procedure is commonly referred to as constant false alarm rate (CFAR).

The interrelation between P_{fa}, P_d, and the signal-to-noise ratio (SNR) can be seen in Figure 4.7, which has been constructed for the single-pulse detection of a non-fluctuating target signal return in Gaussian noise. This figure can be used in a number of ways to determine the parameters in the detection process. If the specifications were given that a P_d of 50% were required with a P_{fa} of 10^{-6}, then the required SNR can be determined. For this case, the SNR would need to be about 11.3 dB. Conversely, if the signal-to-noise ratio was known to be 10 dB, and a P_{fa} of 10^{-6} was required, the P_d can be immediately determined to be about 22%.

The curve in Figure 4.7 is appropriate only for a single pulse for a non-fluctuating target return. For a given situation the same P_d and P_{fa} can be achieved by integrating a number of pulses, each of which has a smaller SNR. Curves are available which show this smaller SNR required for a given P_d and P_{fa} for a variety of number of pulses integrated. The advantages to be gained by using pulse-to-pulse and scan-to-scan integration will be discussed in Chapter 11.

For the situation where the returns from the target are fluctuating, the curve in Figure 4.7 is also not appropriate. Swerling (1954) defined four target fluctuation models and developed curves for signal-to-noise ratios required for given P_d and P_{fa}. Swerling's four cases are:

Fig. 4.7 Probability of detection for a single pulse for a non-fluctuating target (Bailey and Randall, 1961). Reprinted from Electronics, March 17, 1961. Copyright © 1961 McGraw-Hill Inc. All rights reserved.

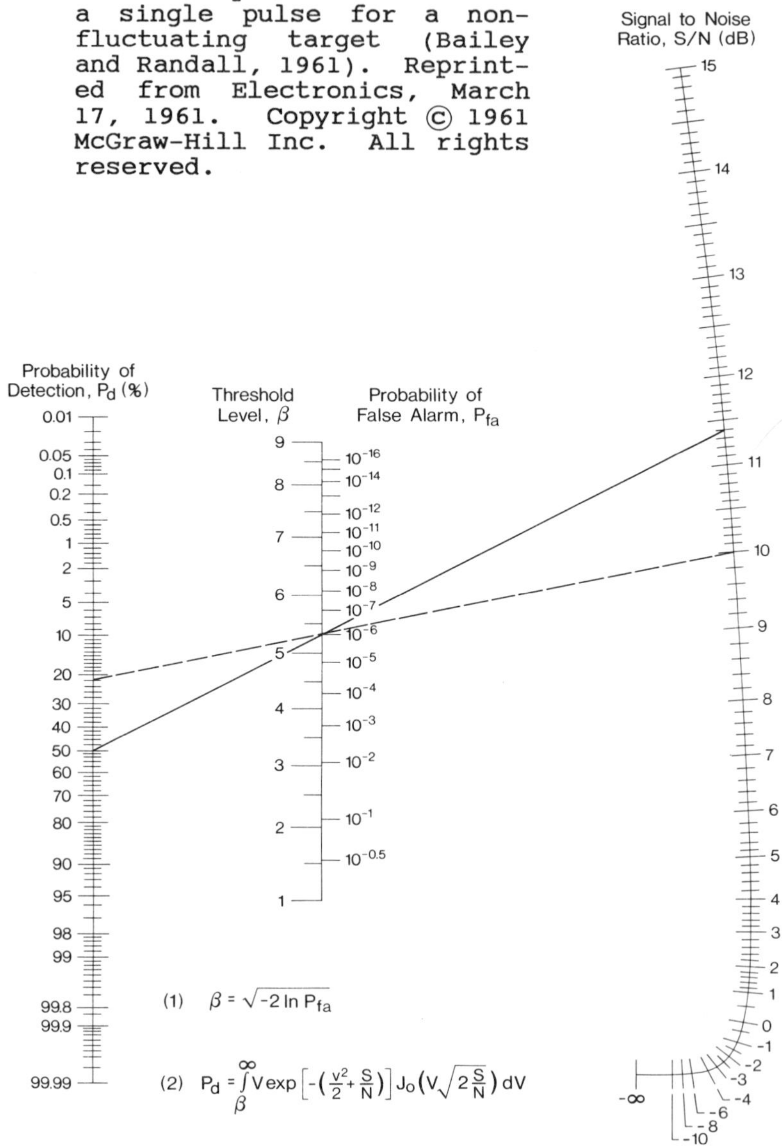

<u>Case 1</u> The returns from the target on a pulse-to-pulse basis are correlated, but the returns from scan-to-scan are independent, and the target consists of many independent scattering elements where no single (or a small number) scatterer predominates.

<u>Case 2</u> The returns from the target on a pulse-to-pulse as well as a scan-to-scan basis are independent, <u>and</u> the target has the same characteristics as for Case 1.

<u>Case 3</u> The returns from the target are as in Case 1 (pulse-to-pulse correlated), <u>but</u> the target consists of one (or a very few) main scattering elements.

<u>Case 4</u> The returns from the target are as in Case 2 (the returns are uncorrelated pulse-to-pulse and scan-to-scan) <u>and</u> the target consists of one (or a very few) main scattering elements.

For the application of radar sensing and detection of ice, Cases 1 and 3 are the ones that apply, as the returns from ice targets are normally correlated on a pulse-to-pulse basis.

<u>Clutter-Limited Case</u>

The more general case in the marine environment is the clutter-limited case where the main source of clutter is sea clutter. Also clutter may result from rain, or, in the case of navigation in ice, ice that is obscuring the target of interest (such as icebergs or another ship). In the case of sea or rain clutter, the returns vary with time and may be described statistically. Sea clutter statistics will be discussed in Chapter 6.

If it is assumed that the signal return from any single scatterer in the clutter is not large compared to the total return from the radar resolution cell, then the set of returns can be considered as conforming with the Central Limit Theorem, yielding a probability distribution that is Gaussian. As in the case of the noise discussed earlier, after video detection this results in a Rayleigh distribution. This is a reasonable assumption for low- to medium-intensity sea clutter, using low-resolution radar systems. As the clutter intensity increases, or the radar resolution is improved, the statistics of the clutter begin to depart substantially from the Rayleigh distribution.

With the assumption of Rayleigh distributed clutter, the same method as used in the noise-limited case can be employed to determine the single-pulse P_d or P_{fa}. Unlike the noise, however, the clutter is correlated from pulse-to-pulse, reducing the effectiveness of integration. As in the noise-limited case, significant improvement in P_d can be achieved through signal processing, as will be discussed in Chapter 11.

Another important consideration in predicting radar detection performance is the effect of the environment on the radar signal propagation path. Chapter 6 will introduce the theory pertaining to these effects.

4.7 SUMMARY

This chapter reviewed basic radar theory, particularly as it applies to surface-based radar systems. Concepts discussed have included the radar equation, radar cross-section, incidence and grazing angles, and the important consideration of local grazing angle. The chapter has also discussed detection range, radar resolution and target detection, including definitions of probability of detection (P_d) and probability of false alarm (P_{fa}).

CHAPTER 5
Experimental Program

As indicated in Chapter 1, one of the serious deficiencies in the use of surface-based radar systems in an ice environment was a lack of understanding of the interaction between surface-transmitted radar signals and ice, together with the experimental data on which to develop this understanding. To overcome this, the Canadian Department of Fisheries and Oceans and the Communications Research Laboratory of McMaster University, Hamilton, Ontario, have been conducting an experimental radar/ice research program since 1981.

The 1981 experiment was the first attempt to systematically evaluate the effects of various radar parameters on the ability to classify and detect ice types and features. During the 1981 program, radars were operated at X- and S-bands. The effects of antenna beamwidth, transmit pulse length, and antenna elevation were evaluated. A limited attempt (using a bistatic radar system) was made to evaluate polarization effects. An experiment in 1982 concentrated exclusively on evaluating the effects of multipath interference over an ice cover. The 1983 experiment was a major multi-faceted program. Radars were operated at S-, X-, K_u-, and K_a-bands. In addition, the radars at X-, K_u-, and K_a-band were operated in a dual-polarized configuration.

The three experiments described above were all conducted during the late spring months when a full and fixed ice cover was present. The 1984 experiment, however, was conducted in the fall during the open-water period. This experiment used a dual-polarized X-band radar to evaluate the effects of polarization on the ability to detect ice targets in open water.

The philosophy throughout the research has been to conduct tightly-controlled experiments with good calibration and detailed surface verification.

5.1 LOCATION

The research program had the objective of developing a better understanding of the interaction of a surface-based radar signal with ice of many types: first-year and multiyear sea ice, pressure ridges, icebergs, bergy bits, and growlers. It was, therefore, necessary to choose an experimental site which gave exposure to all these ice types. Also, to provide good experimental control and surface verification, it was essential that the ice cover remain stationary during the course of the experiment. Finally, in anticipation of future open-water experiments it was important that the landfast ice clear away from the selected location for some period during the summer season.

These conditions are met in just a few areas of the Canadian Arctic, including the entrance to Lancaster Sound (Figure 5.1). Choosing the field site in this area had a further advantage: Lancaster Sound is potentially an area of increased marine activity, thus the experimental work would relate directly to ice conditions in an area where the information is needed.

Logistical considerations were a key to the location chosen on the south side of Lancaster Sound, the most northern point on Baffin Island (Figure 5.1), rather than

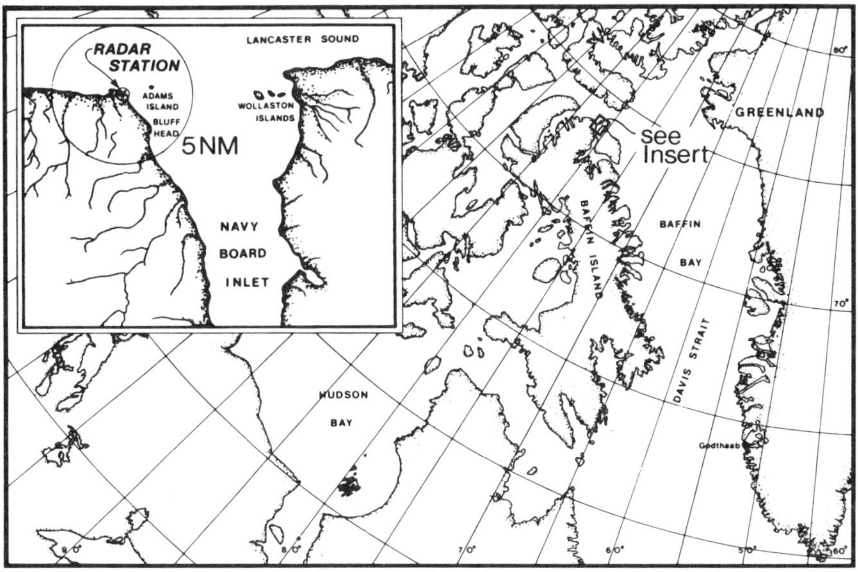

Fig. 5.1 Location of Department of Fisheries and Oceans radar research station on northern Baffin Island.

across Lancaster Sound on Devon Island. A site-selection team surveyed the possible sites on 18 February, 1981 with the following criteria in mind:

- the site should be about the height of a ship's mast (~30 m) above the ice surface;
- the site should overlook stable ice but if possible also a region of dynamic ice;
- the ice should have as much variability as possible; and
- the site must have sufficient flat ground surface to set out a runway for Twin Otter aircraft.

From the pre-trip investigations, a preliminary location on Bluff Head had been selected for further investigation. Upon examination of the ice conditions off Bluff Head (Figure 5.1), however, it became evident that the ice was too smooth and offered little variability. As a result, the search was expanded northward closer to the Navy Board Inlet/Lancaster Sound junction. An almost ideal site was finally located 5.5 km northwest of Bluff Head. The site selected was just over 30 m above sea level. It overlooks the transition zone between Navy Board Inlet where the ice is usually landfast, and more mobile ice in Lancaster Sound. The area also has a good selection of icebergs; most years a number run aground or are frozen into the ice at this location. The location of the selected site is shown in Figure 5.1 at 73°43'21" N, 81°31'02" W.

Shortly after the site was selected, construction began on the camp facilities shown in Figure 5.2. The main building (6 m x 7 m) served as the laboratory and accommodation for the researchers. This building (on the left of the figure) is at the edge of the bluff with no obstructions or ground clutter between it and the ice surface. The second major building (5 m x 7 m) served as

Fig. 5.2 Radar research facilities on northern Baffin Island. Laboratory on the left, accommodation buildings on the right.

galley and food storage. A third building (3.5 m x 5 m) is an accommodation building used by the support crew. A 2.5 m x 3.5 m building was constructed for the generators and for equipment repair. A 400 m x 30 m runway was cleared and marked by barrels and easily handles the Twin Otter aircraft which support the camp.

Finally, three 15-m guyed steel towers were erected to support dipole antennas for communication with Pond Inlet (Baffin Island) and Resolute (Cornwallis Island) and to support meteorological equipment.

For all experiments, field logistical and technical support was provided by the Arctic Research Establishment (ARE) in Pond Inlet. The ARE employs native Inuits who have been trained in scientific and meteorological data collection. During the programs, the ARE personnel collected field samples, took meteorological observations, provided on-ice support, and maintained camp facilities.

5.2 THE 1981 EXPERIMENT

This field experiment was carried out between 24 March and 4 April, 1981. The objectives of the program were to investigate a number of radar parameters to determine their effect on the ability to classify ice type from radar signal returns at low grazing angles. These parameters were:

- the frequency of the radar signal (X- and S-band equipment was evaluated);
- the pulse length of the radar transmission;
- the height of the antenna above mean sea level;
- the horizontal beamwidth of the antenna; and
- the polarization of the returned signal.

Although the final objective of the research program is to be able to classify ice types when both the radar and the ice are moving (as in the case of a ship navigating in ice), to reduce the variables in this experiment and to permit extensive surface verification, the experiment was designed for both stationary targets and stationary signal sources.

The ice conditions consisted of intermittently flat and lightly-ridged ice to the southeast and east, and more heavily-ridged ice to the northeast. To the north was a small area of lightly-ridged ice with heavily-ridged ice at the junction of the landfast ice in Navy Board Inlet and the dynamic ice of Lancaster Sound. In addition, there were a substantial number of icebergs of various sizes within range of the radar. The meteorological conditions were relatively stable during the program, with air temperatures averaging -25°C. The visibility was generally greater than 25 km, while the winds averaged 6 knots.

Three radar systems were available for use during the experiment: a Decca RM 1229 X-band radar; a Decca RMS 1230 S-band radar; and a Furuno FRM 64W X-band radar. The specifications for these radars are given in Appendix B. An RCA Adviser wideband video recorder was used to record the video signals from the radars. The specifications for this unit are also listed in Appendix B. The Decca X-band radar was operated with both a 2.8-m and a 1.2-m antenna, and the Decca S-band radar was operated with a 3.7-m antenna.

The Furuno X-band radar used a 1.8-m antenna in transmit/receive mode and was capable of accepting the input (in a receive-only mode) from a 1-m parabolic dish for the polarization experiment.

The radars were linked to the video recorder through interface units which adjusted the video levels, provided azimuth change pulses (from the scanner position signals), and a heading reference.

In addition to the usual field planning and preparations, calibration of the two Decca radars was carried out, and the complement of antennas was calibrated at the Communications Research Centre (CRC), Ottawa, before shipment to the Arctic.

To determine the actual level of signals received from targets of interest during the field experiment and during subsequent analysis, two calibration procedures were undertaken. The first calibration correlated the level of signal (as viewed on an oscilloscope) at the output of the intermediate frequency (IF) amplifier/ detector of the radars with the received (injected) signal. An on-frequency signal (9,400 MHz or 3,040 MHz) of known amplitude was injected into the front end of the radar receiver. The detection of this signal was monitored on the oscilloscope at the output of the IF amplifier. The input signal was reduced until the detected signal was just visible in the receiver noise. This level of signal was defined as the minimum detectable signal (MDS) of the receiver, and for these radars was between -95 and -96 dBm. The measured MDS agreed favourably with the manufacturer's specification of -96 dBm. The injected signal was then increased in 5-dB steps with the corresponding voltage being measured on the oscilloscope. A plot of IF amplifier output versus received (injected) signal was prepared for each radar. The fast time constant (FTC) and sensitivity time controls (STC) were maintained in an off position and the gain was set to maximum for this calibration.

The second calibration undertaken was to prepare a table of received (injected) signal versus receiver gain setting (measured by gain bias in volts) such that the received signal was just detectable on the plan position indicator (PPI) of the radar. This calibration was accomplished with the same set-up as described previously, except that the oscilloscope was connected to a point in the video amplifier just after the video gain circuit. By adjusting the brilliance on the PPI to a level just below the appearance of speckle (noise) on the display, the appearance of the signal at the monitor point in the video amplifier also caused the signal to appear on the PPI. For ease of calibration, the appearance of the signal at the monitor point was used to provide the calibration. A digital voltmeter was used to measure the bias to the video amplifier.

Starting at an injected signal of -90 dBm, the gain setting necessary to cause a signal to appear on each radar was measured and recorded. The signal was then increased in 5-dB steps and the corresponding bias gain voltages and input signal were again recorded.

At the same time that the MDS and gain voltage calibrations were performed, the output power, pulse width, pulse repetition rate, and pulse shape for the radars were determined. The output power was measured with a peak power meter for long pulse widths, and computed from average power for long, medium, and short pulse widths.

Personnel at CRC undertook antenna calibrations for the Decca X-band, 1.2-m and 2.8-m antennas, for the 1.8-m Furuno X-band antenna, and for the 1-m X-band dish. The S-band antenna was not calibrated as it could not be handled in the small radar range at CRC and because the large radar range was committed to another program.

After a review of the experiments to be carried out and the limitations and capabilities of the radar equipment, the radar antennas were installed as shown in Figure 5.3. The Furuno antenna was mounted on the northeast corner of the radar laboratory (left side of Figure 5.3) at about 34 m above sea level. The antenna was below the peak of the roof but had an unobstructed view of the ice from northwest to southeast. The Decca

Fig. 5.3 Radar antennas for 1981 experiment. Decca S-band on the right, Decca X-band on the scaffolding, and Furuno X-band on front left corner of the laboratory.

S-band antenna (right side of the figure) was installed just below the peak, and almost in the centre of the roof at 35 m above sea level. With the exception of a small amount of shadowing from the scaffolding supporting the Decca X-band radar to the north, the S-band antenna had an unobstructed field of view. The Decca X-band antenna and transceiver were installed on the north side of the building on scaffolding (centre of the figure) to facilitate the variable height experiments. The usual height of the Decca X-band antenna was 36.5 m above sea level, but when this antenna was operated at other heights it was noted in the data. The elevation of each antenna was determined by transporting a precision barometer from sea level to ground level at the radar station.

Fig. 5.4 Radars in laboratory (1981). Decca X- and S-band radars on right. RCA Adviser video recorder on the bench on the left.

Figure 5.4 shows the layout of the Decca S- and X-band radars and the RCA wideband video recorder.

Prior to the first experiment, a very high frequency (VHF) antenna was mounted on the roof and a 10-W portable VHF radio was set up for communication with personnel working on the ice. An audio tape recorder was wired to the audio output and to the squelch of the radio, such that any transmissions to the radar laboratory were recorded automatically.

Data Collection

The experiment began with the identification of two "standard" lines that were selected for the variability of ice along their path. Standard Line 1 ran almost due east from the radar and traversed rough ice near the shore, several pressure ridges at right angles to the line, a relatively large patch of smooth ice, and then alternate areas of pressure ridges, smooth ice, and rubble fields. Standard Line 2 was at a bearing almost due north from the radar and crossed rough ice near the shore, a short area of smooth ice, a considerable distance of light rubble field, followed by heavy rubble at the transition zone between the stable ice of Navy Board Inlet and the mobile Lancaster Sound ice.

Measurements were taken, video recordings were made, and PPI and A-scope photographs were taken for the following configurations:

a) Decca X-band radar with 2.8-m antenna at long, medium, and short pulse lengths at the standard antenna height (36.5 m).
b) Decca S-band radar at long, medium, and short pulse lengths at the standard antenna height (35.0 m).
c) Furuno X-band radar at long and short pulse lengths at the standard antenna height (34.0 m).
d) Decca X-band radar with 1.2-m antenna at long, medium, and short pulse lengths at antenna heights of 29.0, 30.5, 36.5, 38.0, 39.5, and 41.0 m.
e) Decca X-band radar transmitting with 1.2 m antenna at heights of 36.5, 38.0, 39.5, and 41.0 m, with the Furuno radar receiving with regular antenna at standard height of 34.0 m.
f) Decca X-band radar transmitting from 1.2 m antenna at heights at 36.5, 38.0, 39.5, and 41.0 m, with the Furuno radar receiving with parabolic dish in horizontal and vertical polarization at standard height of 34.0 m.

The recordings and photographs were taken for each configuration with both the antenna scanning, and the antenna fixed along the standard lines.

To aid in alignment of the antennas along the standard lines and to provide a target of approximately-known radar cross-section for calibration purposes, Luneburg lenses (targets of known and constant radar cross-section) were deployed at 2.2 and 5.25 nm on Line 1 and at 1.3 nm on Line 2. As a further aid in aligning the radar antennas with the reflectors on the standard lines or any other target of interest, each radar antenna was fitted with a telescope pointing along the centre line of the horizontal beam pattern of the antenna. The telescope was used to quickly sight the lens, then the antenna was gently rocked to maximize the signal on the oscilloscope (A-scope).

Video Recording

Video recordings were made with the RCA Adviser wideband recorder for each configuration, for scanning operation, and for the antenna fixed along Lines 1 and 2. Although the RCA recorder is a two-channel unit, a fault was suspected in video channel B, so only single-channel recordings were made. Consequently, comparative recordings (i.e., between the S- and X-band radars) were made consecutively rather than concurrently, with the time between recordings minimized to eliminate any effects from environmental changes. The Decca radars were linked to the recorder through specially designed interface units that provided the correct video levels, a heading marker, and azimuth change pulses (ACPs) to provide bearing information to the recorder. The interface between the Decca radars and the recorder had the capability of playing the recorded signals back on the radar display to confirm that successful recordings had been obtained. The Furuno radar also had a dedicated interface to the recorder.

For video recording of signals from the standard lines, the antenna was positioned along the standard line by maximizing the signal received from the Luneburg lens, as described in the previous section.

PPI Photography

A large number of photographs were taken of the plan position indicators (PPIs) of the three radars for all the configurations measured. To determine the actual signal return from selected targets during subsequent analysis, the gain of the receiver (which had previously been calibrated) was reduced in 5- or 10-dB steps until no targets were visible on the PPI. Photographs were taken of each gain setting for each parameter.

The photographs of the PPI were taken with a Linhoff 4 x 5 plate camera using Polaroid Type 55 positive/ negative film. The shutter of the camera was held open (in a darkened room) for a number of radar scans until a satisfactory exposure was made. The positive/negative film yielded a positive print to confirm that a satisfactory picture had been taken, plus a negative which could be used for future analysis or enlargement.

During the photographic sequences the FTC and STC were set to zero and the brilliance set to 95% for no blooming. These settings remained constant for all subsequent photographs. A typical photograph of the PPI of the Furuno X-band radar is shown in Figure 5.5.

A-Scope Photographs

When the radar antenna had been pointed precisely along the standard lines by peaking the response from the Luneburg lens on the A-scope, photographs of the A-scope for each of the various configurations were taken. A typical A-scope trace on Line 1 is shown in Figure 5.6.

Polarization Experiment

To evaluate the effects of receiving cross-polarized returns from ice features, the Decca X-band radar (horizontal polarization) was aligned along Line 1 and the received signal was displayed on the lower trace of the

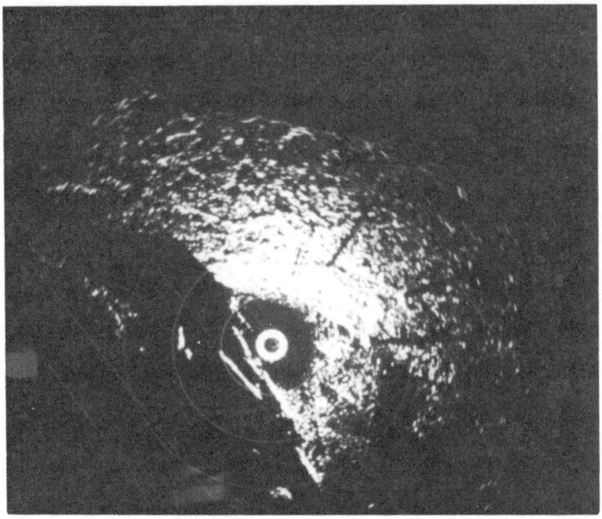

Fig. 5.5 Photograph of Furuno X-band PPI.

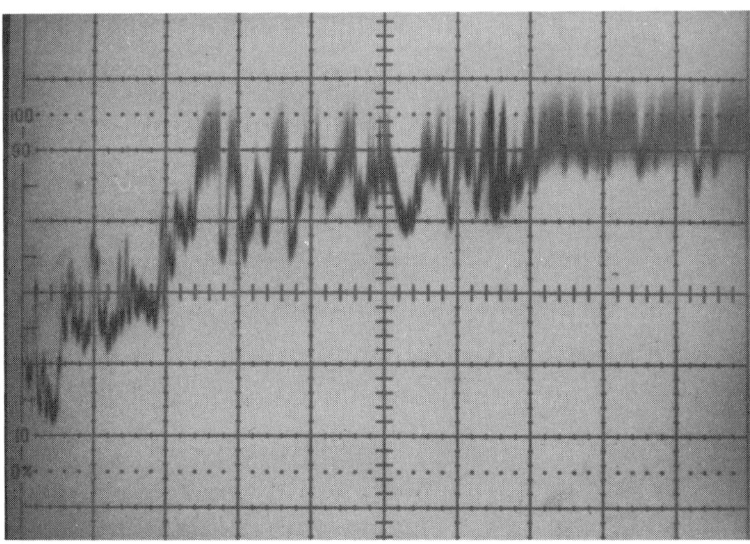

Fig. 5.6 Photograph of A-scope display showing trace along Line 1 (1981).

A-scope. The Furuno radar antenna was replaced by a 1-m parabolic dish antenna which could be rotated for receiving horizontally- or vertically-polarized signals. The parabolic dish was connected to the Furuno radar and was also aligned along Line 1 for maximum output from the Luneburg lens. The output of the Furuno radar in receive-only mode was displayed on the top trace of the A-scope. Photographs of the A-scope were taken for horizontal and vertical polarizations of the parabolic antenna. In addition, the video from the Furuno receiver with the parabolic antenna in both horizontal- and vertical-polarization mode was also recorded on the video recorder.

Figure 5.7 shows the signal received from the Furuno radar with the dish in vertical polarization on the top trace and from the Decca radar with the horizontally-polarized antenna on the bottom trace.

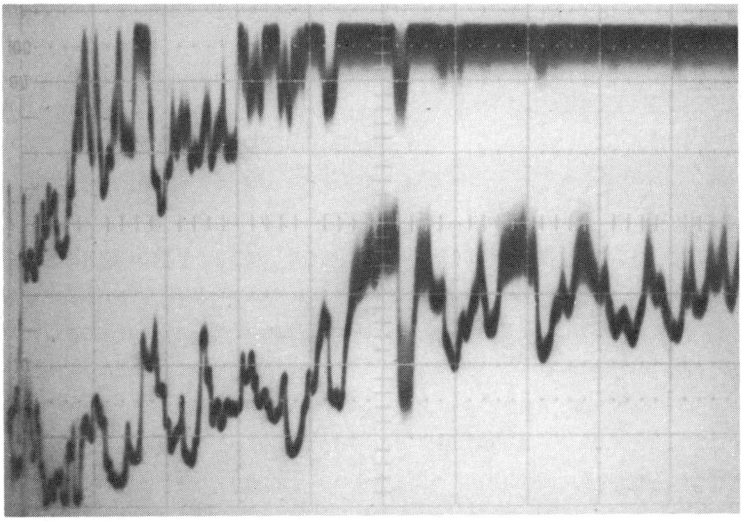

Fig. 5.7 Dual-polarized A-scope trace, with like-polarized X-band return on the bottom and cross-polarized return on the top for Line 1 (1981).

Aerial Photography

Two series of 9 in x 9 in aerial photographs were taken with a Wild RC-10 aerial camera in order to provide a detailed record of the ice conditions in the survey area. This photography was carried out from a Twin Otter aircraft flown from heights of 3,000 and 1,200 m above sea level, yielding scales of 1/20,000 and 1/8,000, respectively. The photography was collected between 1100 and 1600 hours local time, on 30 March, 1981.

Survey markers of strips of cloth (1 m x 3 m) or lamp black were used to mark reference locations on the ice for the aerial photographs. The position of these references was determined by radar fixes. The photographs were assembled into a photo mosaic of the ice surface, as shown in Figure 5.8. Note the large iceberg at A, other smaller icebergs at B, C, D, E, and F, multiyear ice floes at G, H, I, and J, and first-year pressure ridge at K. The mosaic was used extensively in the analysis to correlate radar returns with surface features.

Surface Observations

During the course of the experiment, many trips (with resulting surface descriptions) were made across the ice. Skidoos pulled sleds which carried a radar experimenter, a Luneburg reflector, and an Alpine X-band radar transponder. In some operations the Alpine transponder was mounted on the sled pointing toward the radar station and the progress of the surface crew was monitored on the radar. On other occasions, the location of the surface crew could be monitored by the reflections from the Luneburg lens and the skidoo. While the surface crew progressed across the ice, the radar experimenter provided a recorded commentary of the ice being encountered. When areas of particular interest were encountered the skidoo stopped and surface photographs were taken.

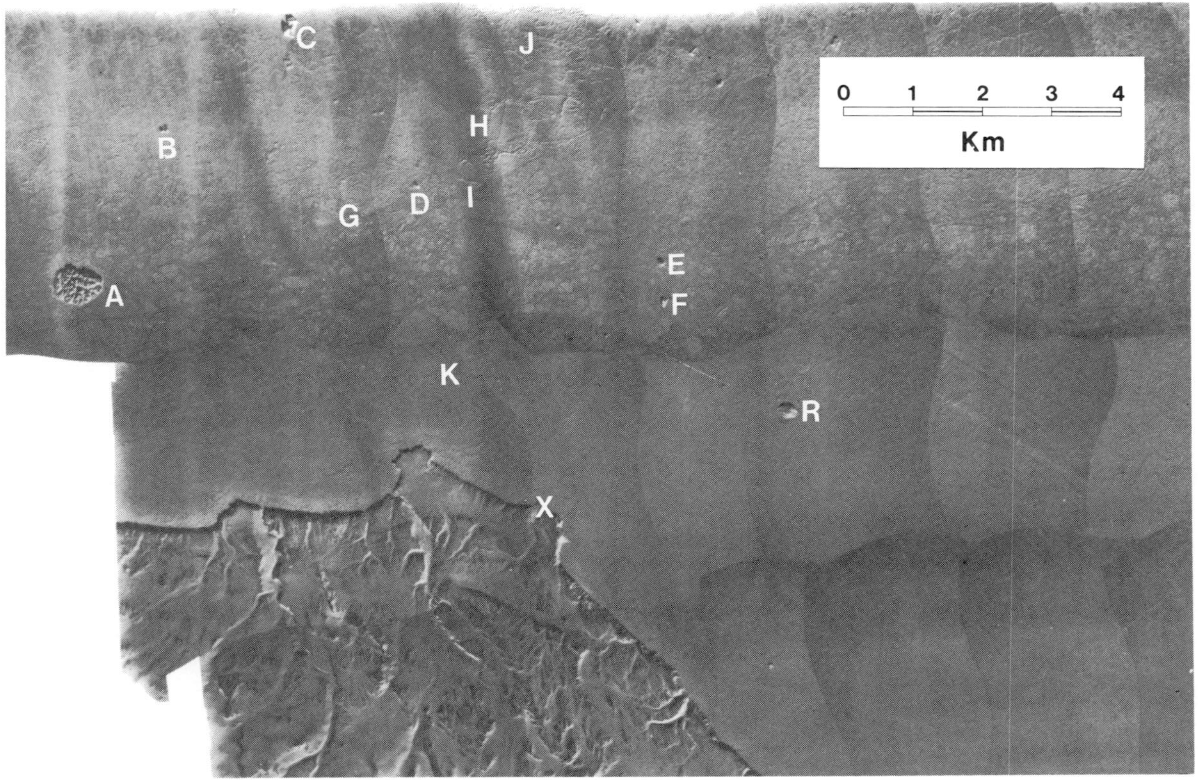

Fig. 5.8 Aerial photographic mosaic of ice conditions during 1981 experiment. Targets of interest include icebergs at A, B, C, D, E, and F; multiyear ice floes at G, H, I, and J; first-year pressure ridges at K. A small island (Adams Island) is located at R and the radar camp is at X.

Standard Line 1 and sections of Line 2 were traversed repeatedly with the skidoos and detailed descriptions of the ice features along them were obtained. In addition, numerous surface photographs were taken along these lines. During the cross-polarization experiment, several significant targets on Line 1 were visited, photographed, and described in detail.

In addition, measurements of surface ice temperature and salinity, ice thickness, snow temperature, snow density, snow salinity, and ice freeboard were taken at a number of locations. Surface ice samples were obtained for analysis at several locations.

5.3 THE 1982 EXPERIMENT

As a result of findings from the 1981 experiment and a general interest in confirming the presence of multipath interference over an ice surface, a dedicated multipath experiment was performed. This experiment was carried out beween 16 and 20 May, 1982. A complete description of the multipath phenomenon is covered in Chapter 6.

The 1982 experiment had two main objectives:

- to measure the radar returns from a target of constant radar cross-section as it was raised from the surface of ice; and
- to measure the power received on the ice as a function of height above the ice surface.

The ice conditions encountered during this program were similar to those for the 1981 program. To the east, and south of Adams Island, the ice was generally smooth, while closer to Adams Island the ice had areas of intermediate roughness interspersed with areas of smooth ice. To the north and northeast at a distance of more than about 2 km from shore the ice was heavily rubbled and ridged. There were four medium-sized icebergs frozen into the ice, north and northeast of the radar camp. The weather was stable during the program with relatively strong winds of about 20 knots. The temperatures averaged -18°C.

In the first experiment, an X-band continuous waveform (CW) signal source (travelling wave tube) with output power of 10 W (40 dBm) was connected to a 1-m dish antenna with a gain of 35 dB. This transmitter was located at the laboratory building situated 30 m above the ice surface. A microwave average-power meter fed by a 22-dB standard gain horn was deployed on the ice (Station 1), at a range of 5,600 m from the laboratory. The measuring system is shown on the right side of Figure 5.9. The ice between the transmitter and the target was relatively smooth, while the ice within 1 to 2 km of the target was very smooth. The grazing angle at the target was less than 0.4°.

Fig. 5.9 Multipath measurement system used in 1982 experiment. Power measurement system (right) and raising rig for Luneburg lens (left).

The transmitting dish antenna was adjusted in azimuth and elevation to give a maximum reading on the power meter. The standard gain horn was then raised from the ice surface to a height of about 2.5 m in 15-cm increments. The received power was measured and recorded for each increment. These measurements were then repeated at eight other locations 3 m apart on a rectangular grid.

The power measurement system was then moved to a new location, Station 2, 3,700 m from the transmitter in an area of rougher ice which extended about 1 km towards the transmitter. This ice consisted of random slabs of ice 0.6 to 1.2 m in thickness, thrown haphazardly up on end with a spacing of 0.6 to 1 m between the pieces. Power measurements from the surface to a height of 2.5 m were repeated at this site on a grid of five locations in an L-shaped pattern, with 3 m spacing between locations.

In the second experiment, a Luneburg lens was deployed on the ice at the 5,600 m site. A standard Furuno 7 kw (68.5 dBm) marine radar was used to illuminate the target.

The radar antenna was stationary, and was carefully aligned on the target to give a maximum return signal from the radar's video amplifier as displayed on an oscilloscope. Once a maximum signal had been received, the radar antenna was locked in that position for the balance of the experiment.

The Luneburg lens then was raised systematically in 15-cm increments from the ice surface to about 4 m using a rope and plastic pipe with no measurable radar cross-section. The return signal at each height was measured and recorded. The raising rig and Luneburg lens are shown on the left side of Figure 5.9. The ladder and power meter were removed for this experiment.

5.4 THE 1983 EXPERIMENT

This major experimental program was carried out between 2 May and 27 May, 1983. The objectives of the experiment were:

- to investigate the effect of radar frequency (over as wide a range as possible) on the ability to differentiate ice types;
- to investigate the use of polarization to differentiate ice types;
- to investigate the use of a coherent radar for ice classification; and
- to conduct further investigations on multipath interference over ice.

To assist in program planning and ice feature identification during the experiment, one month prior to commencement of the field program the ice surface (within a 10-km radius of the radar station) was photographed at a scale of 1/20,000 and 1/8,000, with a high-resolution 9 in x 9 in aerial camera. The 1/20,000 photographs were incorporated into a mosaic of the survey area as shown in Figure 5.10. Features of interest include the icebergs at A, B, C, D, and E, multiyear ice floes at F, G, H, and K; first-year pressure ridges at L and M; smooth first-year ice at O; and rough first-year ice at P and Q. The small island, Adams Island, is located at R.

The ice conditions during the program were similar to those of previous years but there was considerably more multiyear ice present. There were eight icebergs frozen into the ice within radar range from the camp. The temperatures during the program were quite variable, from a high of +1°C to a low of -18°C, with an average of about -10°C. The winds were variable with a maximum of 30 knots. Visibility was generally greater than 10 nm.

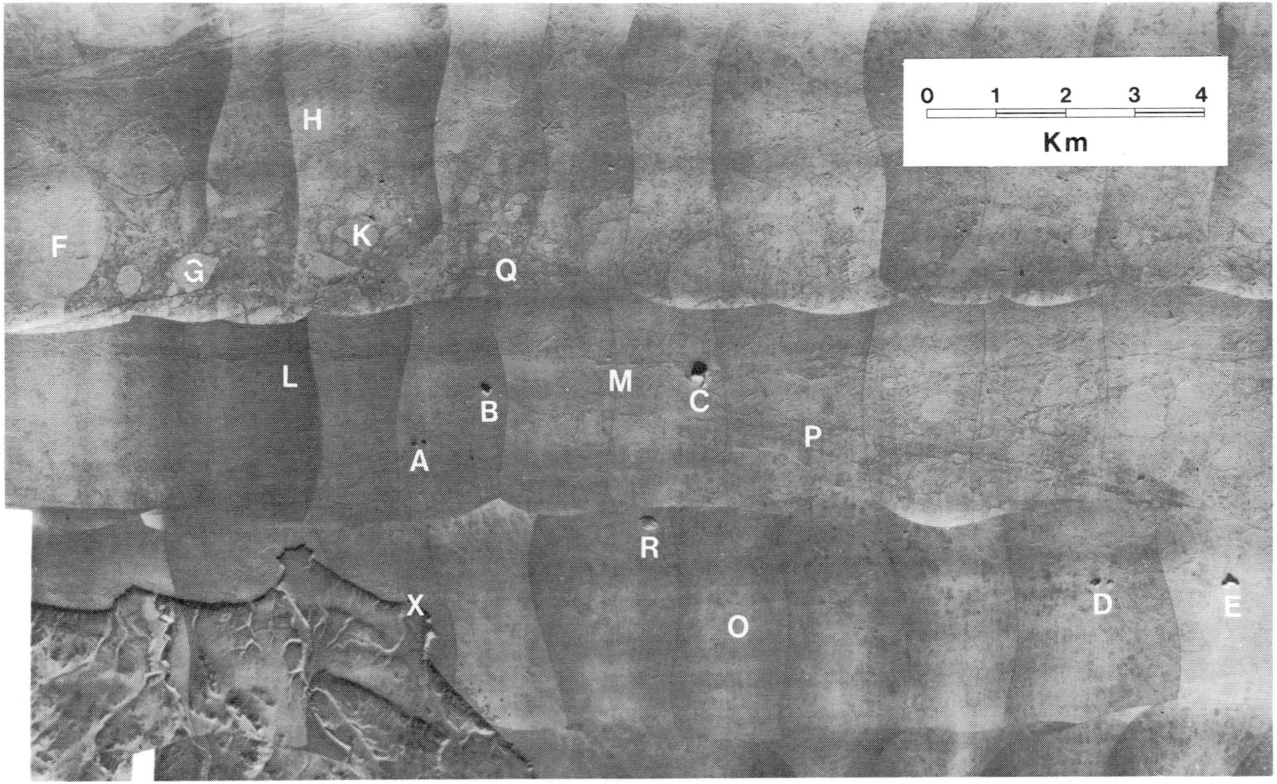

Fig. 5.10 Aerial photographic mosaic of ice during the 1983 experiment. Targets of interest include icebergs at A, B, C, D, and E; multiyear ice floes at F, G, H, and K; first-year pressure ridges at L and M; smooth first-year ice at O; rough first-year ice at P and Q; Adams Island at R and radar camp at X.

To aid in surface investigation of those targets which returned unusual signals, a Bell 206L helicopter was chartered for the last three weeks of the program.

Radar Equipment

A total of seven radar systems operating at S- (3 GHz), X- (9.4 GHz), K_u- (16 GHz), and K_a- (35 GHz) bands were employed during the experiment. Radars at the three higher frequencies were operated in a dual-polarized mode. The radar signals were transmitted in either horizontal or vertical polarization through a dual-polarized parabolic dish antenna with the received like- and cross-polarized returns independently processed and recorded. The S-band radar employed a slotted waveguide antenna, and was capable of transmitting and receiving only in horizontal polarization. From the height at which the radars were operated on the roof of the laboratory (35 to 40 m above the ice), the grazing angle of the radar signal with the ice surface varied between 6.0 and 0.1° over the 300-m to 10-km range of interest. In the prime area of interest from 1 km outward, the grazing angle was less than 2°.

The S-band radar employed was the same Decca RMS 1230 used in the 1981 experiment (see Appendix B for specifications). Two of the X-band radars used were the Furuno FRM 64 and the Decca RM 1229 also employed in the 1981 experiment (Appendix B). A third X-band radar used was the Litton APS-504(V3) airborne radar with 100 kw output power. This radar was operated in conjunction with the Decca X-band radar to give like- and cross-polarized measurements.

The Litton radar fed a dual-polarized parabolic antenna with dual rotary joints mounted on a Decca turning unit. The parallel-polarized waveguide was connected to the Litton radar and the cross-polarized waveguide was connected to the Decca RM 1229, which served as a second receiver. The Decca transmitter was disabled. Specifications for this combined system are given in Appendix B.

The final X-band radar was a low-power coherent radar constructed by the Georgia Institute of Technology, Atlanta, Georgia. Unfortunately, as this radar was being tuned up, the power amplifier failed and the radar could not be repaired. Consequently, no data were collected with this system.

The K_u- and K_a-band radars were dual-polarized radars constructed by the Georgia Institute of Technology. The specifications for these radars are also given in Appendix B. For scanning operation, because only one rotary joint was available, the K_u-band antenna was manually scanned through the field of interest using the one rotary joint and flexible waveguide. For the K_a-band radar, scanning operation was carried out manually through a small sector, using flexible waveguide on both channels. The returns for the K_a-band radar were substantially below the returns for the other radars because of a weak transmitter and lossy waveguide.

Figure 5.11 shows several of the radar antennas on the roof of the laboratory, while Figure 5.12 shows some of the radar systems in the laboratory.

Data Collection

The data were collected in two basic forms: with the radar antennas fixed along one of ten standard lines, and with the radar antenna scanning. The ten standard lines were chosen using the aerial mosaic, aided by surface observations from a helicopter, to incorporate as wide a variety of ice types and features as possible. These lines were chosen to include pressure ridges, smooth and rough multiyear ice floes, large and small icebergs,

Fig. 5.11 Radar antennas used during 1983 experiment. From right to left: Decca S-band, Furuno X-band, K_a-band, and K_u-band.

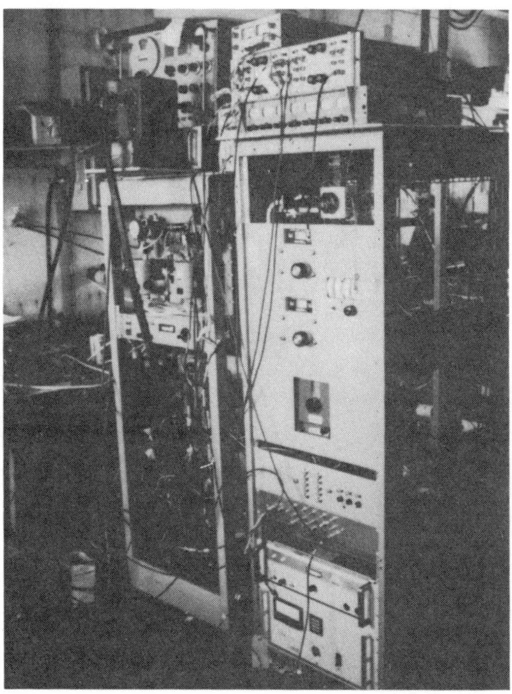

Fig. 5.12 Radar equipment in laboratory (1983); coherent X-band radar and radar sampling system.

heavily-rubbled first-year ice, and smooth first-year ice. The lines were labelled by their bearing with respect to Adams Island, which presented a strong, recognizable return on the radar screen. The standard lines are listed in Table 5.1. To provide repeatable positioning of the antenna, the lines which did not contain large identifiable targets were marked with radar reflectors.

To provide a good description of targets causing particular radar returns, the standard lines were documented extensively. In addition to the aerial photography, a set of oblique photographs was taken from the helicopter at 500-m intervals along each line. Surface photographs were taken of significant targets with a person in each

Table 5.1 Standard lines (1983).

Line number and bearing (degrees) relative to Adams Island	Description of target
0	Adams Island
11*	Corner reflector, diplane, small iceberg
20	Corner reflector, two small icebergs
247	Luneburg lens, multiyear ice floes
252	Luneburg lens, multiyear ice floes
277	Luneburg lens, large rough multiyear ice floe
292	Small iceberg, multiyear ice floes
307	Small iceberg, multiyear ice floes
319	Corner reflector, small and large multiyear ice floes
341	Large iceberg

* The corner reflector and diplane on this radial were used as the primary absolute-calibration targets.

scene for scale. Finally, snow samples were obtained at several locations and ice cores were collected from 14 suspected multiyear floes.

When collecting radar data along the standard lines, the antenna was rocked carefully back and forth until a maximum signal was received from the reference target, either reflector or ice feature, then locked in this position. For data collection with the antenna scanning, the radar video as well as azimuthal information and north-up marker were recorded independently.

The radar data for each radar were collected using two separate and distinct recording systems. For the first system, the raw radar video, heading marker, and azimuthal change pulses were recorded on the RCA Adviser dual-channel wideband (6 MHz) video recorder (for specification see Appendix B) which had a dynamic range of about 36 dB. The second system used a range-gated digitizer supplied by the Georgia Institute of Technology. This system consisted of a 12-bit, analogue-to-digital converter and a computer controlled range-gate sampler. The sampling system started at minimum range and stepped out in 12-m (80-ns) steps, and could be programmed to collect a number of samples at each range prior to stepping on to the next range. A total of 1,024 range samples could be collected from a minimum range of between 500 and 1,500 m out to a maximum of 10.7 km from the minimum range. In normal operation, 128 samples were taken at each range and averaged to produce a smoothed sample. When operating in antenna scanning mode, the sampling system collected data at the same range at each azimuthal position for one scan, then stepped out to the next range.

As well as facilitating data collection, the data acquisition system enabled preliminary processing and analysis of the data in the field. The characteristics of the data reduction system are shown in Appendix B. Besides giving a comparison of calibrations from day to day, the system allowed examination of range plots. Targets that produced unusual returns could then be examined to determine the cause of the return. Figure 5.13 shows a typical range profile printed in the field (on the top) with the corresponding A-scope trace,

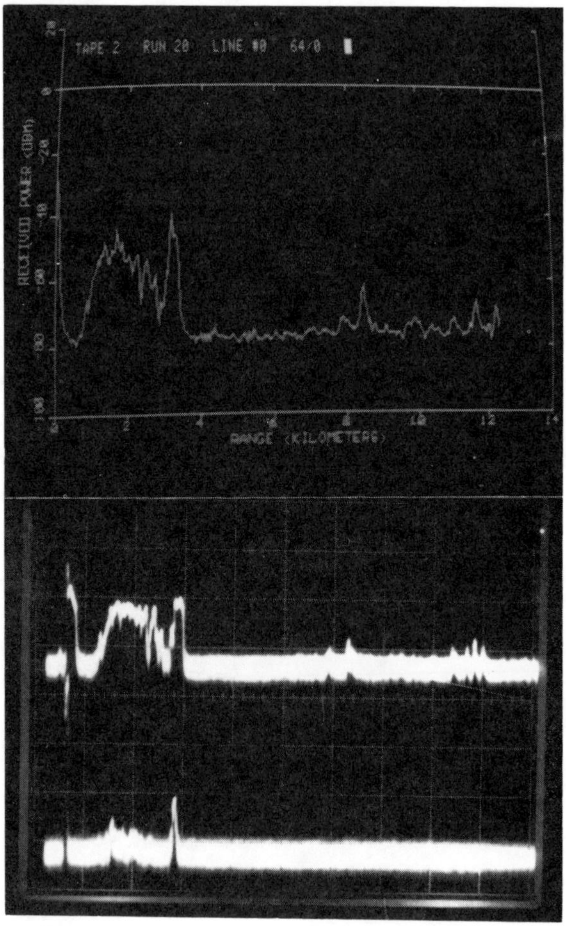

Fig. 5.13 Radar range profile printed in the field (top) and A-scope photo of same returns (bottom).

at the same scale, on the bottom. Reviewing the results in the field also enhanced confidence in the integrity of the data.

To provide meaningful and comparable results, each radar system was calibrated twice daily. Using a signal source at the appropriate frequency, a signal was injected into the front end of the radar, starting from saturation and decreasing to minimum detectable signal. The radar response at each injected level was recorded by the two data collection systems described above. For the Georgia Institute of Technology data acquisition system, the output of the A/D converter at 5-dB input power levels was fitted with a piecewise linear curve to yield an input power to output A/D value transfer curve. As a cross-comparison on this method, radar targets with known radar cross-section were placed at known ranges on Line 11. These targets were a corner reflector with a radar cross-section (RCS) of 35 m^2 at 9.4 GHz, and a diplane with an RCS of 28 m^2 at 9.4 GHz. The diplane seam was tilted at 45° relative to the horizontal to return all the power in the cross-polarized channel. Given the known height of the radar targets, values of the RCS were calculated for each of the targets at each frequency, taking into account multipath interference. The measured values closed to within 3 to 4 dB of the calculated values, which is considered good in an experimental RCS program. In addition to the calibrations performed, signal losses in the waveguide runs to the antennas, pulse width, pulse repetition frequency, and output power of the radars were measured regularly.

Multipath Experiment

A multipath experiment was also carried out during the 1983 program. A corner reflector with RCS of 36 m^2 at X-band was raised from the ice surface to a height of 4 m in 15-cm increments. The raising rig was the same as shown in Figure 5.9 but with the corner reflector replacing the Luneburg lens. The reflector was illuminated by an X-band radar with the return signal power recorded for each height increment. The reflector was aligned in azimuth at each height to give a maximum return. The experiment was performed in an area of smooth ice at a range of 4,800 m and in an area of rough ice at a range of 3,700 m.

5.5 THE 1984 EXPERIMENT

This open-water experiment was conducted between 11 and 23 September, 1984. The objectives of the experiment were:

- to investigate the depolarizing characteristics of icebergs, bergy bits, and growlers in open water; and
- to investigate the depolarizing characteristics of sea clutter.

The ice conditions consisted of open water with nine grounded icebergs and a continually changing suite of small sea-ice and glacial-ice targets. During the experiment, one large ice floe moved through the experimental area, but otherwise the sea was open, with less than one-tenth cover by floating ice pieces. The temperatures were in the range of -2 to -8°C. The winds were light with a maximum recorded speed of 12 knots which unfortunately resulted in low sea states, thus limiting the sea clutter data that could be collected.

The radar system employed consisted of two identical Furuno FR-1011 marine radar transceivers operated through a dual-polarized dish antenna. The antenna was a 1.2-m parabolic antenna with a 2.2° beamwidth and a gain of 38.5 dB. The antenna had a polarization isolation greater than 25 dB. The radars operated at X-band (9,420 MHz) with an output power of 7 kW on medium pulse

(530 ns). Specifications for the radars are shown in Appendix B. The minimum detectable signals for the two receivers were about -96 dBm and the radar PRF was 1,600 Hz. The radar transmitters were synchronized and, by appropriate disabling of the magnetrons, allowed operation with simultaneous horizontal transmit/horizontal receive (HH) and horizontal transmit/vertical receive (HV), or vertical transmit/vertical receive (VV) and vertical transmit/horizontal receive (VH).

The antenna was sighted manually on an ice target using a telescopic sight and aligned to give maximum radar return from the target. Data werethen collected for two to three minutes. Between data collection runs, the antenna was re-sighted if necessary, so that the target was always in the main lobe of the antenna beam.

The radar video was recorded on the RCA Adviser video recorder; it was also digitized and recorded with a specially-designed Video Data Acquisition System (VDAS) with specifications as shown in Appendix B. The VDAS system consisted of an 8-bit microprocessor-controlled front-end digitizer, feeding an IBM PC microcomputer controller-logger. The digitizing and buffering systems were configured to take 64 samples in range for 128 successive sweeps, at a sampling rate of 5 MHz. The data was then transferred to a 10-megabyte hard disc on the IBM PC and later was archived onto floppy diskettes.

One problem encountered in the data collection was the limited dynamic range of the radar receivers for a given gain setting. Consequently, three radar gain settings were required to provide the full dynamic range of interest.

The ice targets were measured for various durations at intervals from a few minutes for some of the faster-moving small targets, to five days for the large stationary icebergs. They were sampled at a maximum rate of 128 successive sweeps over a duration of 0.08 s. For longer durations, multiples of such sample sets were taken, or the PRF clock fed to the sampling system was reduced by a factor between 1 and 512, resulting in the collection of 128 sweeps decimated in time.

Surface verification of the targets was provided by following the iceberg targets by boat and making sketches, notes, and taking photographs from four aspects with a calibration target in view for reference. The boat moved well outside the antenna beam during data recording. The target parameters collected included length, width, height, surface ice conditions, wetness of surface, interaction with the sea surface, and aspect toward radar during recording. Sea conditions were noted including wave direction, height, and period. Meteorological conditions taken were air and sea temperature, wind speed and direction, humidity, precipitation, and cloud cover. Before each data run a photograph of the target, using a telephoto lens, was taken from the roof of the laboratory.

During the course of the experiment, a total of thirty-three ice targets were measured. The size of the targets varied from that of medium-sized icebergs (3,200 m^2) to growlers of less than 1 m^2 surface area. Each day, the radar receivers were calibrated using injected signals, and the transmitter power, pulse length, and pulse shape were measured, together with waveguide losses.

5.6 SUMMARY

This chapter described a series of four ice radar experiments carried out at a research station in the Canadian Arctic. The analysis of the data collected during these experiments is presented in Chapters 6, 7 and 8.

CHAPTER 6
Environmental Effects

The amount of radar energy returned from a target of interest and the target's detectability are strongly influenced by several environmental factors. The pattern propagation factor (mainly caused by multipath interference), absorption of energy in the atmosphere, and anomalous propagation conditions can significantly affect the level of energy returned from a target. In addition, returns from the sea (sea clutter) can limit the detection of targets in open water.

This chapter discusses these four environmental factors and their influence on radar detection.

6.1 PATTERN PROPAGATION FACTOR

The pattern propagation factor takes into account non-free-space propagation conditions caused by multipath interference.

As shown in Figure 6.1, multipath interference (also known as the Lloyd's Mirror Effect) occurs when a direct signal transmitted from a radar near the earth's surface is interfered with by a signal reflected off the surface. This well-known phenomenon over the sea surface will, in this work, be shown to occur over a smooth ice surface as well. Multipath interference results in

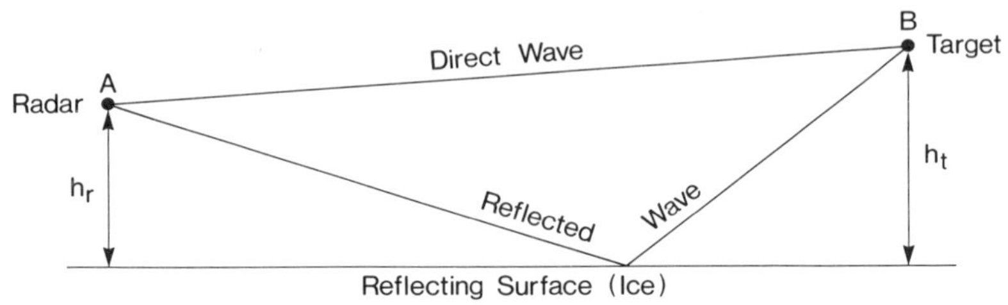

Fig. 6.1 Diagram of multipath interference.

destructive or constructive interference, depending on the phase between the direct and reflected signals.

The radar equation (Equation 4.1) was for a signal propagating in free space, where the propagation factor F, is equal to one. However, for propagation over a plane reflecting surface (earth, sea, ice, etc.), F has to be modified to take multipath interference into account. This results in the modified radar equation:

$$P_r = \frac{P_t G^2 \lambda^2}{(4\pi)^3 R^4 L} \sigma_t \cdot 16 \sin^4\left(\frac{2\pi h_a h_t}{\lambda R}\right) \qquad (6.1)$$

where the returned power is now modulated by the term

$$F^4 = 16 \sin^4\left(\frac{2\pi h_a h_t}{\lambda R}\right) \qquad (6.2)$$

which cycles from a maximum of 16 (in decibels, 12 dB) to a minimum of zero.

For small angles where $\sin X \simeq X$, Equation 6.1 can be reduced to:

$$P_r = \frac{4\pi P_t G^2 \sigma_t (h_a h_t)^4}{\lambda^2 R^8 L} \qquad (6.3)$$

For a point target, the power returned falls off as $1/R^4$ when the effect of multipath interference is ignored. However, when the modulating term as shown in

Equation 6.2 is added, the return power drops off at the $1/R^4$ rate only out to what is known as the transition range that is defined as follows:

$$R_T = \frac{4h_a h_t}{\lambda} \qquad (6.4)$$

after which the power drops off as $1/R^8$.

Within the interference region, that is the area from the transmitter out to the transition range, the return power is modulated by Equation 6.2 such that the power experiences successive peaks and nulls as shown in Figure 6.2. The locations of the peaks and nulls are determined by the height of the antenna, the height of the target, and the wavelength of the transmitted signal.

The effect of multipath interference at a particular range can be visualized by plotting the modulating lobes

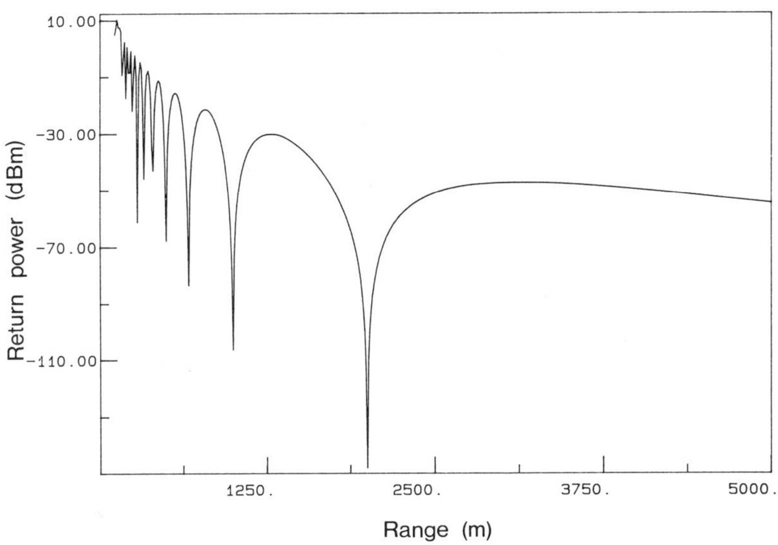

Fig. 6.2 Antenna pattern lobes in range caused by multipath interference.

as a function of the height of the target above the surface. Figure 6.3 is a plot of the first three interference lobes for an X-band (3 cm) radar operating at a height of 30 m and range of 5,600 m. It can be seen that the lobe maxima occur at 1.4, 4.2, and 7.0 m, and the minima occur at 2.8 and 5.6 m. Equations 6.1 and 6.3 are based on the assumption of a perfectly reflecting surface. As this is seldom the case, the theoretical maxima will not be realized and the nulls (minima) will be partially filled in.

Equation 6.3 shows that by increasing the height of the antenna (h_a) and decreasing the wavelength (λ) of the transmitted signal (increasing its frequency), the power returned from a given low-lying target can be increased.

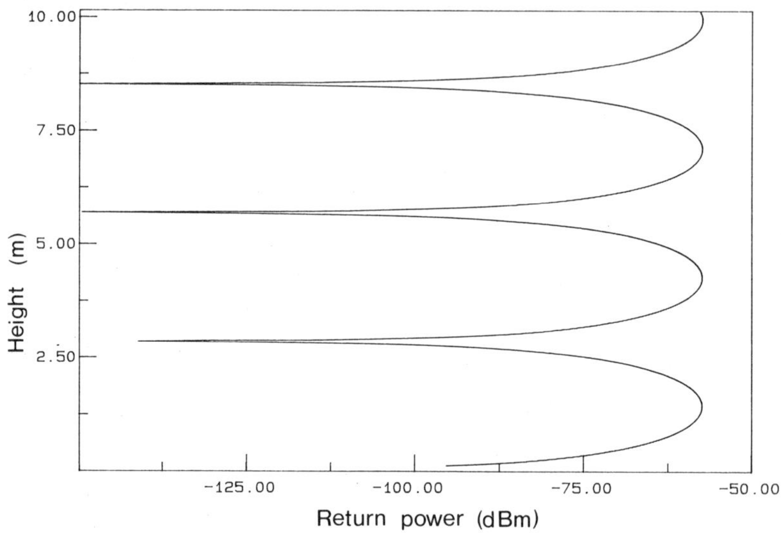

Fig. 6.3 Multipath interference lobes versus height, at range of 5,600 m for an X-band radar with an antenna height of 30 m.

Experimental Results

The results of the 1982 and 1983 multipath interference experiments are presented below. In Figure 6.4, the results of the 1982 on-ice, one-way power measurement at Station 1 are shown. The solid curve is a plot of the predicted power using a modified form of Equation 6.1 for a one-way transmission path, while the solid line at -30 dBm is the power predicted by the free-space radar equation. The dashed curve is a plot of the actual power received as a function of height above the surface. Note that because of the one-way transmission path, the maximum expected signal level increase over the free-space prediction was 6 dB, rather than 12 dB.

As Figure 6.4 shows, the actual signal maximum was exactly that predicted by Equation 6.1. The maximum and

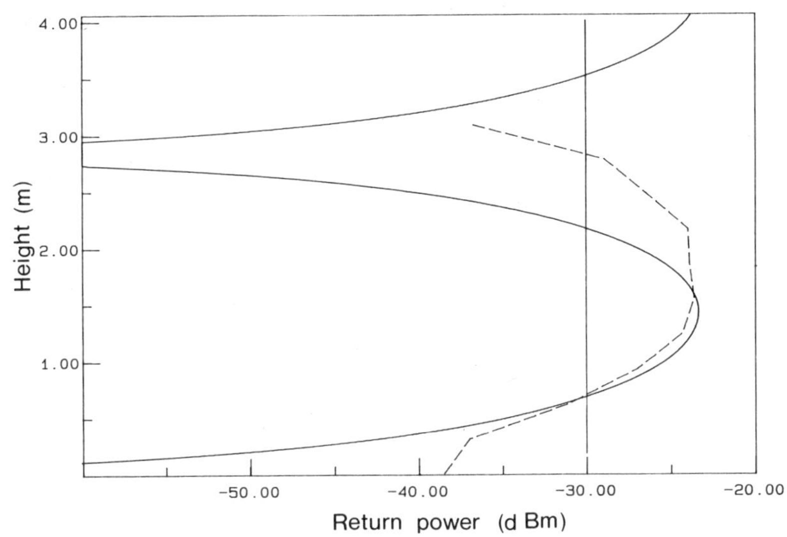

Fig. 6.4 Experimental and theoretical multipath lobes at Station 1 (1982). Results for one-way multipath interference at X-band at a range of 5,600 m. Solid trace is theoretical lobes, dashed trace is experimental results. Solid line at -30 dBm is predicted free space return.

minimum occurred at slightly greater heights than anticipated; otherwise, there is good agreement on the shape of the curves.

In Figure 6.5, the results of the power measurement from 1982 at Station 2 (3,700 m from the transmitter) are shown. This site was in an area of rougher ice, and had rougher ice between it and the transmitter. The dashed curve is a plot of the actual power received as a function of height, while the solid curve again represents the power as estimated by Equation 6.1. The solid line at -26 dBm is the free-space power prediction. Again, there is excellent agreement between the actual and predicted power levels, with the maximum having a value of 5.8 dB greater than the free-space prediction. The null, however, has been reduced to only 3 dB below the

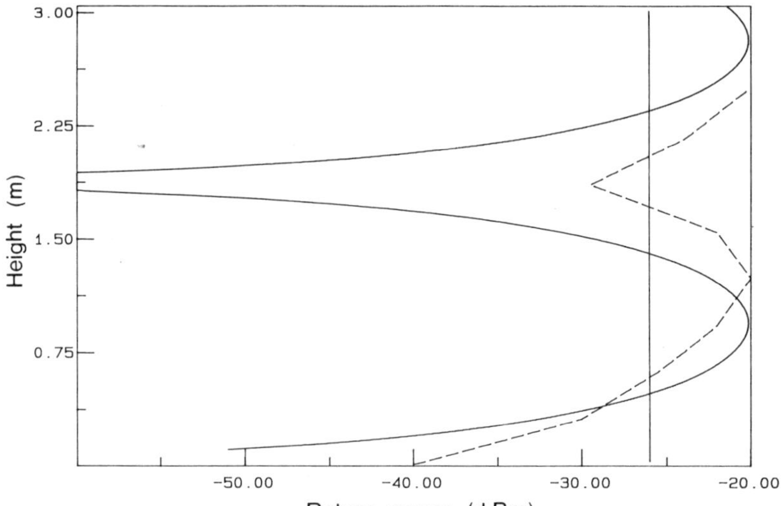

Fig. 6.5 Experimental and theoretical multipath lobes at Station 2 (1982). Results for one-way multipath interference at X-band at a range of 3,700 m. Solid trace is theoretical lobes, dashed trace is experimental results. Solid line at -26 dBm is predicted free space return.

free-space prediction, undoubtedly caused by the rougher ice at this location. As in the case of the measurement at the smooth site, the peak signal at this site also occurred at a greater height than predicted, although the null occurred where predicted.

Figure 6.6 shows the results of the lens-raising experiment (1982) at Station 1. The solid curve is a plot of the signal level as a function of height, as predicted by Equation 6.1, while the solid line at -72 dBm is the return power predicted by the free-space radar equation. The dashed curve is a plot of the return power from the Luneburg lens as a function of the height of the lens above the surface. Although the lens had a constant

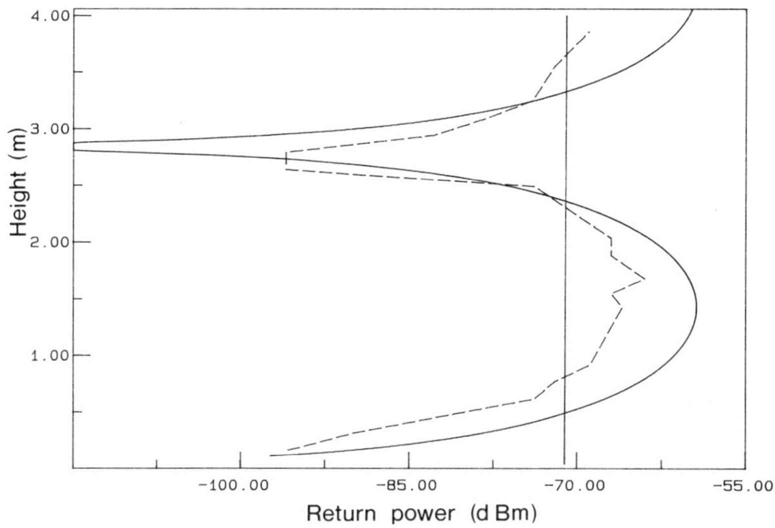

Fig. 6.6 Experimental and theoretical multipath lobes for Luneburg lens at Station 1 (1982). Results for two-way multipath interference at X-band at a range of 5,600 m. Solid trace is theoretical lobes, dashed trace is experimental results. Solid line at -72 dBm is predicted free space value.

radar cross-section, the return power varied over more than 30 dB as the lens was raised from the surface to a height of 4 m. The measured return power agrees favourably with theoretical predictions of the modified equation and deviates substantially from the free-space prediction. The maximum signal level received was only 7 dB stronger than the free-space prediction, instead of 12 dB as predicted by Equation 6.1, while the null was a substantial 25 dB down, but not zero. However, as indicated earlier, the maxima and minima are predicted on the basis of a perfect reflecting surface, which the ice is not.

The results from the 1983 experiment which involved the use of a corner reflector are similar. In Figure 6.7, the returned power from the corner reflector at the

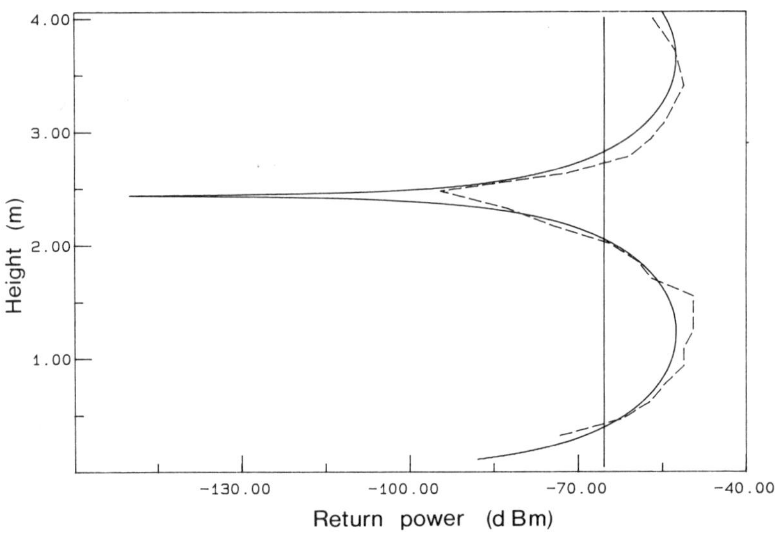

Fig. 6.7 Experimental and theoretical multipath lobes at Station 1 (1983). Results for two-way multipath interference at X-band at a range of 4,800 m in an area of smooth ice. Solid trace is theoretical curve, dashed trace is experimental results. Solid line at -64 dBm is predicted free space value.

smooth ice site is plotted in the dashed trace, while the solid trace is a plot of the predicted signal level as a function of height. There is excellent correlation between the predicted and actual return. In this case the measured level of the peaks is slightly greater than predicted. Figure 6.8 shows the results from the site in the rough ice. The predicted nulls and peaks were almost completely eliminated by the diffuse reflection from the rougher ice. If the multipath interference had been completely eliminated, the return power would have been uniform with height at the free-space value. There are still some peaks and nulls in the return, with the maximum return 3 dB above the free-space value.

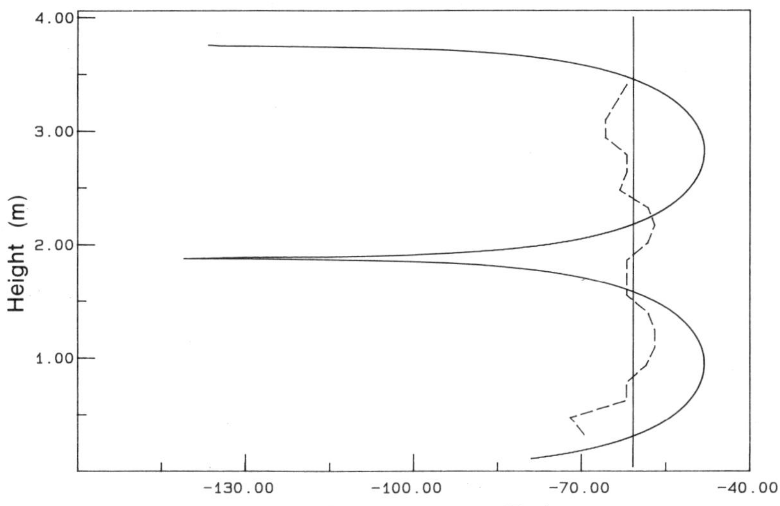

Fig. 6.8 Experimental and theoretical multipath lobes at Station 2 (1983). Results for two-way multipath interference at X-band at a range of 3,700 m in an area of rough ice. Solid trace is theoretical return, dashed line at -60 dBm is predicted free space value.

Discussion

The experimental results show that multipath interference, particularly in smooth ice areas, has a significant effect on the level of radar signals received over the ice surface. From these results it can be seen that the application of the free-space radar equation for estimation of the detectability of targets near the ice surface will result in significant errors. The modified radar equation (Equation 6.1), on the other hand, provides a good estimate of the returned energy or the energy received on the ice.

It is interesting to note that the power received at both Station 1 (5,600 m) and Station 2 (3,700 m) during the 1982 experiment was influenced considerably by multipath interference, even though the 3,700 m station was in, and preceded by, rougher ice. It is estimated that at the near 0.5° grazing angle for this station, ice features of the order of 0.5 m would appear rough to the radar signal (see Rayleigh roughness - Chapter 7), and thus should have broken up the multipath interference. Other than the more substantial infilling of the null, this did not appear to be the case. However, the variation amongst the five measurements taken at the station in the rough ice was substantially greater than that for the smooth ice station, possibly indicating that the rough ice was beginning to break up the multipath interference. The results from the 1983 experiment at the rough ice station clearly showed that the multipath interference had been disrupted.

It is also interesting to note that, although the one-way transmissions in 1982 and the two-way transmissions in 1983 had a peak signal that agreed with the predicted level to within 1 to 3 dB, the two-way transmission experiment with the Luneburg lens resulted in a maximum signal 7 dB less than that predicted. This result may have been caused by the broad angular response of the Luneburg lens and interference from other azimuthal paths.

Referring again to Equation 6.3 (the approximation to Equation 6.1), it can be seen that the power received at the transmitting site is directly proportional to the height (h_a) of the transmitting antenna and inversely proportional to the wavelength (λ) of the transmitted signal. The results of the experiment show that with a transmitter height of 30 m and a range of 5,600 m, signal levels begin to fall below the free-space prediction at about 0.75 m. However, as Figure 6.9 shows, if the height of the radar antenna is lowered to 10 m, the first lobe maximum moves up significantly to 4.2 m. Signals from targets at a height of 2 m or less will begin to fall off rapidly below the free-space prediction.

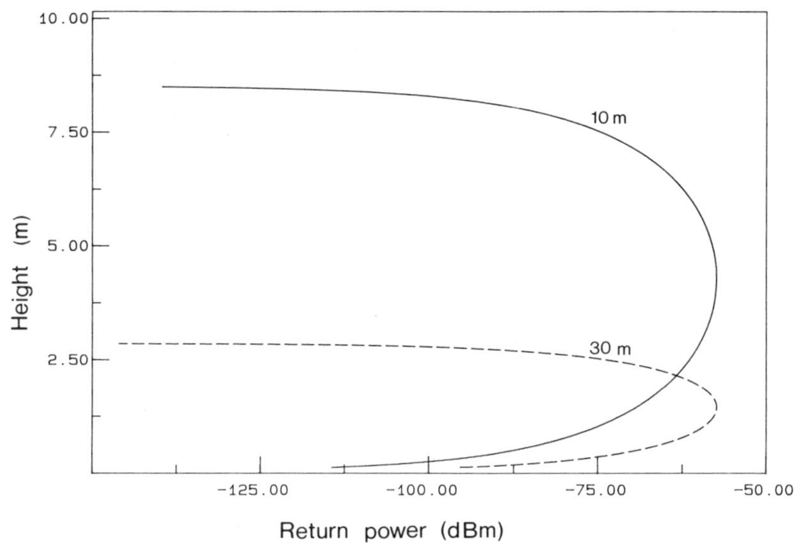

Fig. 6.9 Effect of lowering antenna on multipath lobes. First multipath lobe for X-band radar at a range of 5,600 m. Solid trace is for antenna height of 10 m, dashed trace for height of 30 m.

The power received at the transmitter is also inversely proportional to wavelength; hence, it is directly proportional to the frequency of the transmitted signal. Figure 6.10 shows that as the frequency is increased (S-band at 3 GHz, X-band at 9.5 GHz, K_a-band at 35 GHz), the height at which the signal return from a target falls below the free-space prediction is correspondingly reduced.

In summary, there are three points that warrant emphasis:

1) When concerned with the detection of targets near the ice surface, the radar antenna should be as high as possible, and the radar frequency should be as high as practical, bearing in mind that radar frequencies above X-band suffer increasing atmospheric attenuation.

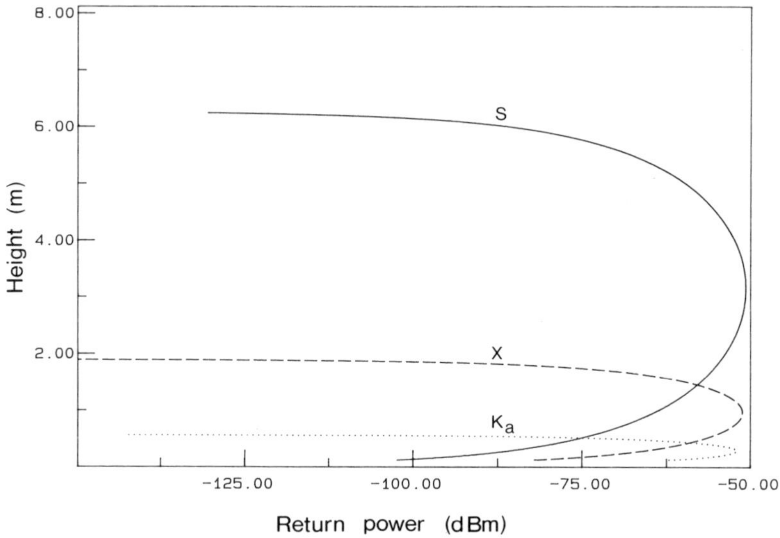

Fig. 6.10 Effect of changing frequency on multipath lobes. First multipath lobes for three frequencies of radars at range of 5,600 m with radar at 30 m. Solid trace for S-band, dashed trace for X-band and dotted trace for K_a-band.

2) When performing calculations to determine the detectability of targets near the ice surface, it is essential that the modified radar equation (Equation 6.1) be employed.
3) When performing radar evaluation or detection experiments with artificial or real targets, care must be taken in positioning of the targets above the ice surface to avoid significant errors from the multipath phenomenon.

6.2 ATTENUATION BY ATMOSPHERIC CONDITIONS

The presence of precipitation, water vapour, and even oxygen in the atmosphere, tends to attenuate and scatter microwave signals transmitted through the atmosphere. For a clear atmosphere without precipitation, signals at microwave frequencies are attenuated by the presence of oxygen and water vapour. Figure 6.11 shows a plot of attenuation due to oxygen and water vapour as a function of frequency. With the exception of the water vapour absorption at 22.246 GHz, which should be avoided, frequencies below 40 GHz should not be adversely affected by attenuation from atmospheric gases.

For polar and temperate regions, Saxton and Hopkins (1951) quote typical attenuation values for atmospheric gases at 3 cm (X-band) of between 0.01 and 0.02 dB/km.

The attenuation caused by fog at 18°C is shown in Figure 6.12. The attenuation in fog is related to the optical visibility, with the attenuation in heavy fog (visibility 100 ft) at 10 GHz (X-band) at 18°C being 0.1 dB/km. Data from Saxton and Hopkins (1951) for polar regions (0°C) for the same foggy conditions (heavy fog - less than 30 m visibility) show an attenuation of 0.2 dB/km. The attenuation by fog in the colder (polar) regions is twice that for more temperate climates.

The attenuation resulting from rainfall is directly related to the rate of precipitation. Table 6.1 from

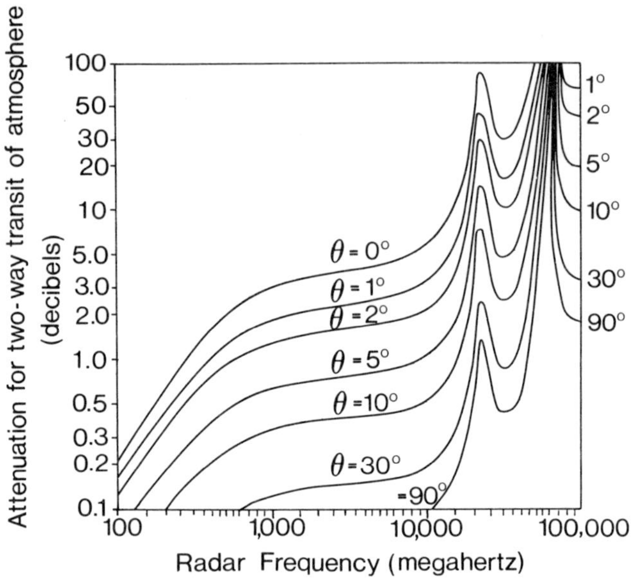

Fig. 6.11 Atmospheric absorption as a function of frequency for several elevation angles (Blake, 1969).

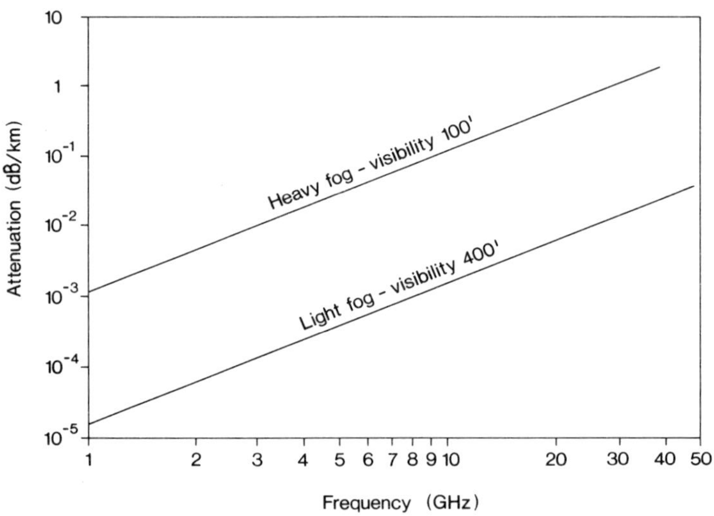

Fig. 6.12 One-way attenuation caused by fog versus frequency. Heavy fog $\bar{M}$ = 2.3 g/m³; light fog $\bar{M}$ = 0.032 g/m³ (Skolnik, 1962, Introduction to Radar Systems, McGraw-Hill).

Table 6.1 Attenuation caused by rainfall.

	Attenuation (dB/km)							
	Polar regions*				Temperate regions**			
Rate of rainfall (mm/h)	When λ=1.25cm	When λ=2.25cm	When λ=3.20cm	When λ=10cm	When λ=1.25cm	When λ=2.25cm	When λ=3.20cm	When λ=10cm
0.5	0.05	0.015	0.005	0.0004	0.04	0.015	0.004	0.0002
1.0	0.09	0.03	0.01	0.0007	0.10	0.03	0.01	0.0004
2.5	0.25	0.08	0.025	0.0015	0.30	0.09	0.03	0.001
5.0	0.55	0.18	0.06	0.003	0.65	0.25	0.07	0.002
12.5	1.50	0.50	0.15	0.007	0.80	0.65	0.23	0.004
25.0	3.1	1.0	0.35	0.015	3.7	1.4	0.55	0.008
50.0	6.5	2.2	0.8	0.03	7.5	2.9	1.3	0.02

* Temperature about 0°C.
** Temperature about 15°C.

(Saxton and Hopkins, 1951, Reprinted by permission of the Institute of Electrical Engineers from the Proceedings of the IEE, Vol. 98, Part III, 1951, p. 26.)

Saxton and Hopkins (1951) shows the attenuation in dB/km for various rates of rainfall at several wavelengths for polar and temperate regions. There is only a marginal difference between the polar and temperate regions, with the attenuation in the temperate regions generally being slightly higher.

The attenuation caused by snowfall is not covered nearly as well in the literature as the attenuation by fog or rain. Saxton and Hopkins (1951) note that the attenuation from ice crystals in the atmosphere, whether they be in the form of hail or snow, is much less than the attenuation from rain of an equivalent precipitation rate. Blake (1980) noted, however, that wet snow can produce a very large attenuation, exceeding even that of an equivalent rainfall, and quoted Bell (1967) who reported a one-way loss of 25 dB at X-band over a 55-km path with 2 mm/hr of wet snow. This observation yields an attenuation rate of 0.45 dB/km. Even if the 2 mm/hr of snow equated to 2 mm/hr of rainfall, which it surely would not, the equivalent rainfall attenuation would only be between 0.02 and 0.03 dB/km. It is evident, then, that wet snow can have a significant attenuating effect on microwave signals.

In addition to attenuating the radar return from the target of interest, the precipitation may also generate its own radar return. If the target is in an area of precipitation, the return from the precipitation may mask the return from the target. For further details see Skolnik (1970).

6.3 ANOMALOUS PROPAGATION CONDITIONS

Variations in the refractive index of the atmosphere can have significant effects on the propagation of radar signals near the earth's surface. The refractive index (n) is defined as the ratio of the velocity of an electromagnetic wave in a vacuum to its velocity in the atmosphere. For a "standard" atmosphere, the refractive index decreases exponentially with increasing height. Consequently, at low altitudes it decreases nearly linearly with height. The refractive index varies from an earth's surface value of approximately 1.0003 to 1 at high altitude.

A more common method of expressing the refractive nature of the atmosphere is in N-units (normally referred to as refractivity) where

$$N = (n-1) \times 10^6 \qquad (6.5)$$

which yields a value of about 300 at the earth's surface. A generally accepted formula for N is (Rotheram, 1982):

$$N = 77.6 \frac{P}{T} + 3.73 \times 10^5 \frac{e}{T^2} \qquad (6.6)$$

where P = pressure (mbar)
T = temperature (°K)
e = water vapour pressure (mbar).

Another method of expressing refractivity is the modified refractivity, M, that causes the earth's surface to be flat by introducing a curvature term in the refractivity profile:

$$M = N + \frac{h}{a} \times 10^6 \qquad (6.7)$$

where a is the earth's radius and h is the height above the surface in km. Using a value of $a = 6.4 \times 10^3$ km, Equation 6.7 can be reduced to:

$$M = N + 157\ h \qquad (6.8)$$

With a normally refractive atmosphere (where N decreases at 39 N-units/km or M increases by 117 M-units/km) near the surface, radio waves refract towards the earth's surface but with a radius less than that of the earth. Standard conditions are considered to exist when the N-gradient is between -79 and 0 N-units/km. Anomalous propagation conditions exist when there are abnormal vertical distributions in the refractive index. If the N-gradient is between -156 and -79, the radio waves will be bent downward at a rate greater than standard but less than the earth's curvature. This situation is known as super-refraction and results in extended propagation range. If the gradient is less than -157 N-units/km the radio waves are refracted downward with a radius exceeding that of the earth, and the waves can become trapped in a channel or duct. This ducting phenomenon has the effect of greatly extending the range of propagation. If the N-gradient is greater than zero, subrefraction occurs and the radio waves bend less, thus reducing propagation range. The ranges of N and M values and the associated propagation conditions are shown in Table 6.2.

Table 6.2 Refractive index for selected propagation conditions.

Condition	N-gradient (N/km)	M-gradient (M/km)
Trapping	$dN/dh \leq -157$	$dM/dh \leq 0$
Super-refractive	$-157 < dN/dh \leq -79$	$0 < dM/dh \leq 78$
Standard	$-79 < dN/dh \leq 0$	$78 < dM/dh \leq 157$
Subrefractive	$dN/dh > 0$	$dM/dh > 157$

(From Hitney *et al*. 1985)

As noted by Hitney and Richter (1976) "ducting phenomena are particularly dramatic and frequent in the marine environment." In the marine environment there are three principal types of ducts: surface-based, elevated, and evaporation ducts. These ducts are shown diagrammatically in Figure 6.13.

Surface-based and elevated ducts are caused by advection of warm, dry air over moist, cold marine air, creating layers of different refractive index. An elevated duct (Figure 6.13a) occurs where a minimum occurs in the M profile. The bottom of the duct is the height where the M value below the minimum is the same as the minimum. Where this second minimum does not occur, a surface-based duct is formed as shown in Figure 6.13b. Thus the condition for a surface-based duct to form is that the M value at the top of the duct must be less than the value at the surface. The advected, surface-based duct has a refractive index that varies with the distance offshore. As the distance offshore increases, the strength of the duct decreases.

Evaporation ducts are shallow ducts that occur frequently over most oceanic areas. The evaporation duct is a result of the saturated air at the sea surface decreasing to non-saturated air above the sea. The decrease in water vapour is roughly logarithmic, giving a refractivity near the sea surface that is modified logarithmically. Figure 6.13c shows the relative humidity and M profile for an evaporation duct where the evaporation duct height is the height at which the M curve reaches a minimum value.

The evaporation duct height is determined from the difference between the air and sea temperature, the wind speed, and the relative humidity. Figure 6.14 is a plot of duct height as a function of these factors. The solid lines labelled 0 to 100% are plots of duct height at wind speed of 1 m/s for various humidities, while the dotted lines are for wind speed of 10 m/s. For evaporation

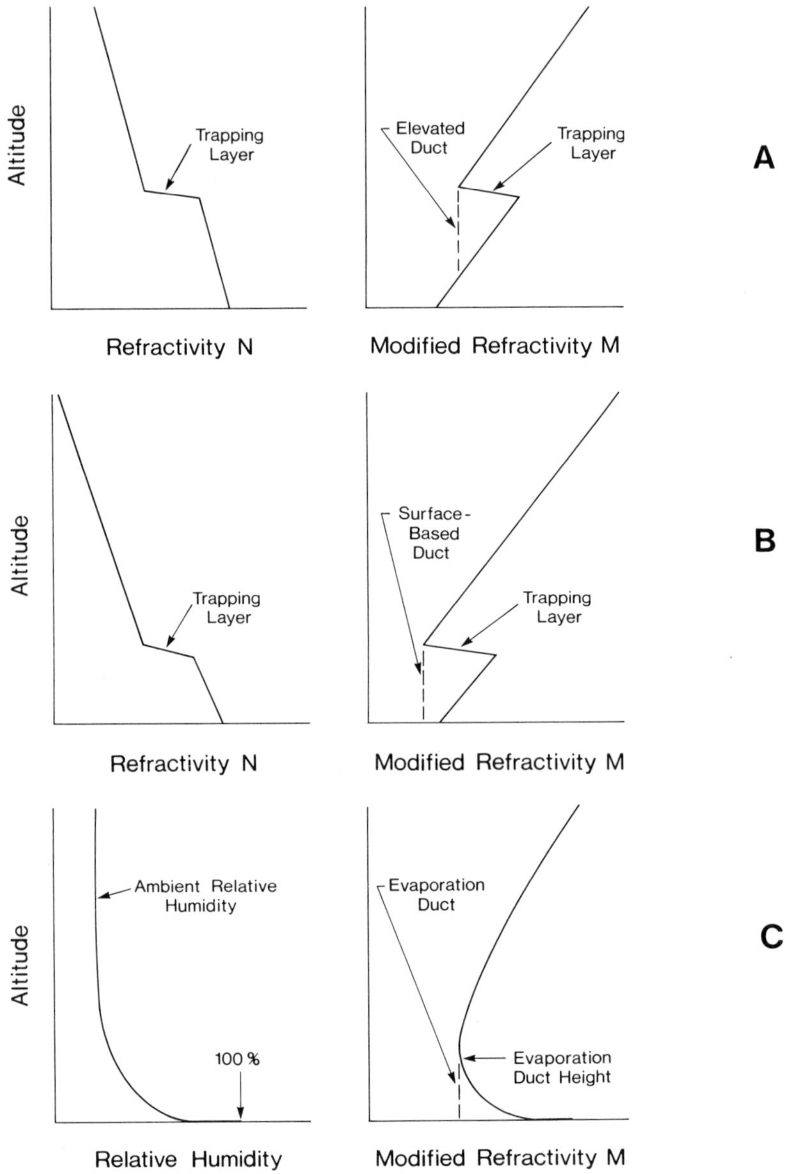

Fig. 6.13 Refractive profiles for atmospheric ducts; elevated, surface, and evaporation (Hitney et al., 1985, Tropospheric Radio Propagation Assessment, Proceedings of the IEEE, Vol. 72, No. 2. Copyright © 1985 IEEE).

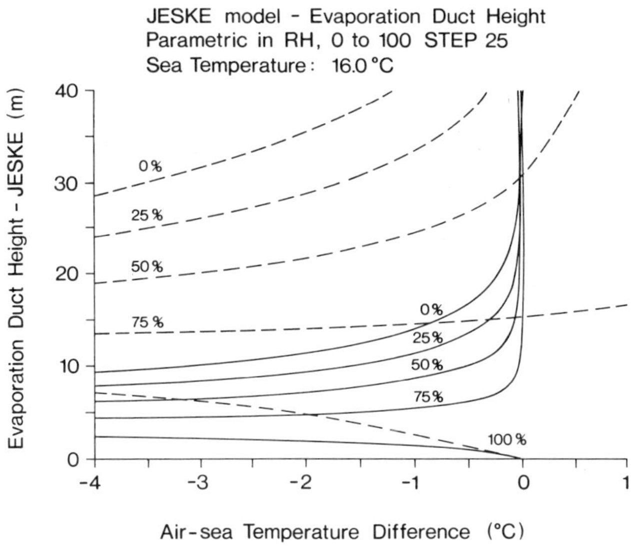

Fig. 6.14 Evaporation duct heights: as a function of air-sea temperature difference, humidity, and wind speed. Solid lines are calculated for wind speed of 1 m/s. Dotted lines for wind speed of 10 m/s (Anderson, 1983).

ducts at northern latitudes, Hitney _et al_. (1985) quote a long-term mean of 8 m, but varying between 0 and 40 m. However, evaporation ducts are very "leaky" and thus affect propagation above as well as within the duct. Evaporation ducts have a lower frequency effect limit of about 3 GHz.

The effect of the duct, in addition to refracting the signal more, is to allow propagation of the signal with little attenuation. Dougherty and Hart (1979) showed the loss, L, for a one-way path between a transmitter and a receiver both located in the duct as follows:

$$L \geq 92.45 + 20 \log f + 10 \log d + 0.03d + A, \text{ (dB)} \quad (6.9)$$

where f is the frequency in GHz, d is path length in km, and A is coupling loss. Dougherty and Hart go on to note that the loss within the duct can be less than the free-space loss (L_{fs}):

$$L_{fs} = 92.45 + 20 \log f + 20 \log d, \text{ (db)} \qquad (6.10)$$

In the duct, the spherical wave spreading associated with free-space propagation is constrained in the vertical direction because of the duct. This results in the power density varying as the inverse distance, not the inverse squared distance. Note the term 20 log d in Equation 6.10, which is only 10 log d in Equation 6.9.

The height of the antenna with respect to the height of the duct is important for maximizing the propagation within the duct. However, as shown by Richter and Hitney (1975) the antenna does not need to be in the evaporation duct to benefit from signal enhancements of the duct. They show results of an experiment conducted off Greece where antennas were mounted at 5 and 20 m; it was assumed that the 20-m antenna was above the duct. When strong ducting conditions occurred, the lower antenna received stronger signals; however, the higher antenna still received much stronger signals than for non-ducting conditions. For the remainder of the time, the higher antenna received stronger signals.

There has been a great deal of work done on modelling of electromagnetic (EM) fields in ducting conditions (Marcus, 1982; Rotheram, 1982; Baumgartner _et al_., 1983; Hitney _et al_., 1985). The models usually employ optical (ray trace) techniques to calculate field strength within the horizon and multimode waveguide techniques beyond the horizon. "Waveguide analysis treats the EM field as one or more discrete families, or modes, of plane waves propagating in a confined region" (Anderson, 1983). The refractive index is approximated by a number of linear sections and there is assumed to be no horizontal variability. Full mode solutions, which can include a hundred or more modes, involve extensive computation. A number of approximation models have been proposed which, in general, consider only a single or a few propagation modes. As noted by Anderson (1983), the modes are numbered by the occurrence of when the mode maximizes its effect to produce a minimum path loss.

Figure 6.15 shows the multimode path loss versus duct height at 18 GHz. The propagation modes are noted on the minimum loss positions. The dashed traces show two approximations using single-mode calculations. The single-mode calculation provides a good approximation up to the first-mode minimum, but then deviates substantially from the full solution. The modified single-mode approximation retains the loss associated with the

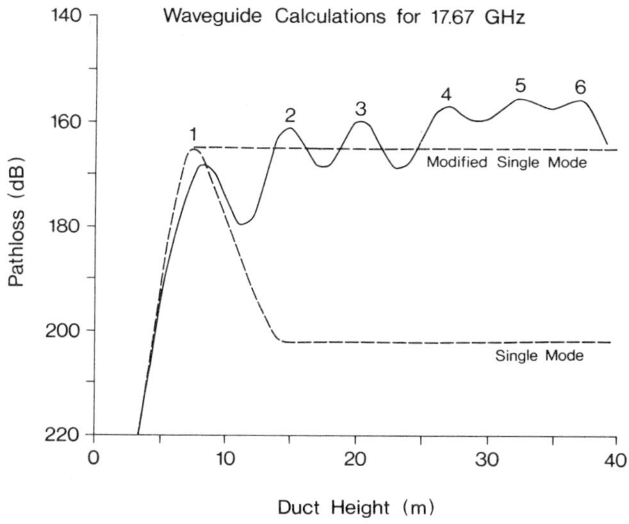

Fig. 6.15 Path loss as a function of duct height, illustrating the distinction between the waveguide, single mode, and modified single mode propagation models (Anderson, 1983).

first-mode minimum above the height at which it occurs, thus yielding a much better approximation. The integrated refractive effects prediction system (IREPS), which is under development by Baumgartner *et al*. (1983), uses a simplified model that "gives quite good agreement with measurements, especially for the more prevalent moderate duct heights for which single-mode solutions are adequate." This comment indicates that for the near surface applications with shallow duct heights, such as considered in this program, single-mode models would be adequate.

The prediction of anomalous propagation conditions is not only limited by modelling difficulties, but also by the reliability and accuracy of the measured meteorological data. Helvey (1983) shows that predictions of ducting conditions based on radiosonde data are significantly overstated, particularly as a result of sensor lags, pre-release exposure to direct sunlight at the surface, and data processing techniques. In particular, the critical humidity parameter suffers from sensor response lags and temperature effects.

Almond and Clarke (1983) also comment on limitations of radiosonde data for propagation prediction, summarizing that "radiosonde carried by a balloon can confirm the presence of a near normal atmosphere or the potential for ducting near the location of the sonde ascent." For accurate prediction more sophisticated systems like airborne refractometers are needed.

However, the height of, and therefore the existence of, an evaporation duct can be determined readily by measuring the sea temperature, and the air temperature, humidity, and wind speed at a fixed height above the ocean surface, and by performing several relatively simple calculations. The equations required to determine the duct height are contained in Hitney (1975).

For maritime areas, surface-based and evaporation ducts will lead to detection of targets at greater range than under normal conditions. Surface-based ducts decrease in strength as the distance offshore increases, while evaporation ducts appear to be a regular occurrence over most sea areas. Since ducting and super-refraction increase the range of detection, thus aiding the offshore operator, the requirement appears to be to detect standard or substandard propagation conditions in order to adjust downward the detection prediction system.

However, for areas off the east coast of Canada, the optimistic expectation that evaporation ducting will occur regularly and thus provide enhanced radar propagation is ill-founded. An analysis by the authors of over 22,000 three-hourly meteorological data records collected from stationary platforms in the Hibernia area off Newfoundland over a five-year period indicates that, for many months of the year, evaporation ducting has a low rate of occurrence. In addition, the regular occurrence of fog and precipitation gives a high occurrence of subrefractive conditions.

In fact, during the heavy fog months of May, June, and July, when visibility is the poorest and need for radar the greatest, the atmospheric propagation conditions are the worst. Figure 6.16 shows a plot of the percentage occurrence of ducting and subrefractive conditions versus the month, for 1984. It is clear that subrefractive conditions have a high frequency of occurrence during the May, June, July period, whereas ducting conditions only occur with a high frequency during the September, October period, which is also characterized by clear visibility. Figure 6.17, which is a plot of percentage of occurrence of ducting conditions for the five years studied, shows that the presence of ducting conditions is relatively consistent. Figure 6.18, which shows the occurrence of

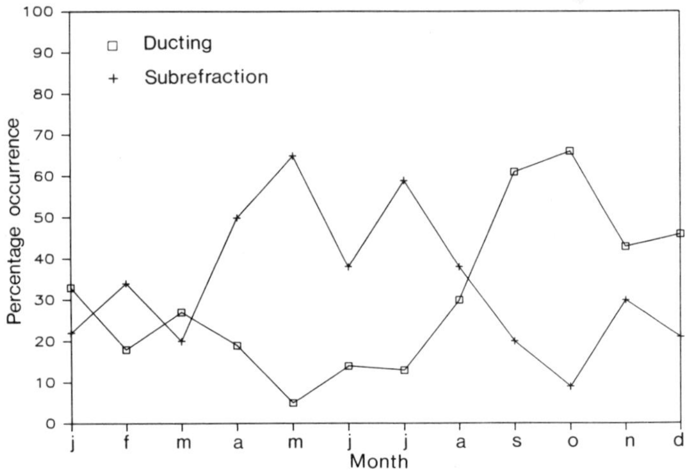

Fig. 6.16 Occurrence of ducting and subrefractive conditions (1984) for offshore Newfoundland.

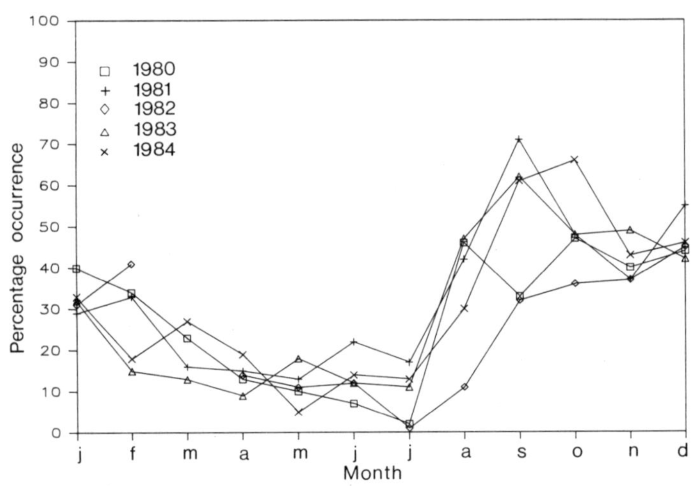

Fig. 6.17 Occurrence of ducting conditions (1980-1984) for offshore Newfoundland.

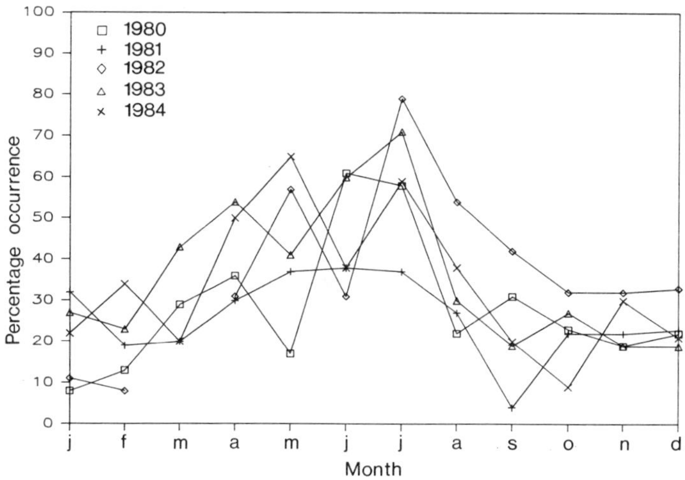

Fig. 6.18 Occurrence of subrefractive conditions (1980-1984) for offshore Newfoundland.

subrefractive conditions, has a larger scatter in the data but does show that subrefractive conditions have a high occurrence during the months when the radar is most needed.

The assessment of the occurrence of ducting and subrefraction was accomplished by running the data through the U.S. Naval Ocean Systems Center propagation model of Hitney (1975).

It has not been possible to assess the severity of degradation of radar propagation during the subrefractive periods. However, operators in the Canadian east coast offshore area should be warned that propagation conditions will be below normal during significant portions of the year.

6.4 SEA CLUTTER

Radar returns or clutter from the sea surface are the most serious impediment to the detection of small targets on the ocean. The severity of sea clutter depends on many factors, including the nature of the sea surface. The condition of the surface of the ocean is the result of several independent processes. Sea swell, the long, near-sinusoidal periodic waves that propagate through the ocean, is caused by distant winds. On top of the swell is superimposed the sea, which consists of relatively steep, short-crested waves produced and driven by local winds. The swell and sea are termed long and short gravity waves, respectively, and are affected by the wind velocity, pressure, and gravity, but not surface tension. Capillary waves are small ripples on the surface of the sea which are caused by local wind and are governed by surface tension. The capillary waves respond instantaneously to the wind and quickly die out when wind velocities are less than 3 to 5 m/s.

The condition of the surface of the sea is often defined by wave height, which is the vertical distance between the top of a wave crest and the bottom of an adjacent trough. A commonly used term, the significant wave height ($H_{1/3}$) is the average crest-to-trough height of the highest one-third of all the waves, while the average wave height is the average of all the waves. The horizontal distance between wave crests is called the wavelength.

The characteristics of the surface of the sea are often summarized in a single term called sea state. However, this term can be misleading since the relationship between wave height, wind velocity, and sea state is not clearly defined. There are two sea-state scales in general use: the Beaufort and the Douglas scales. The Beaufort scale is based on wind velocity and not sea disturbance. The corresponding wave heights are also a function of the length of time that the wind has been blowing and the length of sea (fetch) over which the wind has blown. The Douglas scale is related directly to wave height and is thus more descriptive of the sea surface condition. These two scales have been combined in a common table as shown in Table 6.3.

As shown earlier in this chapter, because of the effect of the propagation factor (multipath interference), the return from a constant cross-section target falls off as $1/R^4$ out to the transition range, then as $1/R^8$ beyond the transition range. Further, as shown in Chapter 4, the radar return from a distributed target falls off at a rate of $1/R^3$ for free-space propagation. When the distributed target (in this case sea clutter) is on the surface it is affected by the propagation factor such that the return falls off as $1/R^3$ out to the transition range, and as $1/R^7$ beyond this range. For sea clutter, the transition range can be determined from (Katzin, 1957):

$$R_T \simeq \frac{5h_a H}{\lambda} \qquad (6.11)$$

where h_a is the radar height and H is the height of the top of the waves. The $1/R^3$ and $1/R^7$ dependency is shown in Figure 6.19.

The radar returns from the sea are dependent on the sea conditions and the radar parameters. The sea conditions include the sea state (wave height) and wind speed, while the radar parameters include grazing angle, radar wavelength, polarization, and the azimuthal position of the radar beam with respect to the sea, i.e., upwind, downwind, cross-wind.

Modern scattering theory attributes radar backscatter from the sea to small facets (particularly on the short gravity waves) and to the capillary waves with lengths comparable to the radar wavelength. The facets are small patches on the gravity waves that act like flat-plate reflectors of various sizes and orientations. The

Table 6.3. Wind and wave scale.

WIND VELOCITY KNOTS	4, 5, 6, 7, 8, 9, 10, 20, 30, 40, 50, 60, 70
BEAUFORT WIND AND DESCRIPTION	1 LIGHT AIR; 2 LIGHT BREEZE; 3 GENTLE BREEZE; 4 MODERATE BREEZE; 5 FRESH BREEZE; 6 STRONG BREEZE; 7 MODE RATE GALE; 8 FRESH GALE; 9 STRONG GALE; 10 WHOLE GALE; 11 STORM
REQUIRED FETCH IN MILES — FETCH IS THE NUMBER OF MILES A GIVEN WIND HAS BEEN BLOWING OVER OPEN WATER	50, 100, 200, 300, 400, 500, 600, 700
REQUIRED WIND DURATION IN HOURS — DURATION IS THE TIME A GIVEN WIND HAS BEEN BLOWING OVER OPEN WATER	5, 20, 25, 30, 35
IF THE FETCH AND DURATION ARE AS GREAT AS INDICATED ABOVE, THE FOLLOWING WAVE CONDITIONS WILL EXIST. WAVE HEIGHTS MAY BE UP TO 10% GREATER IF FETCH AND DURATION ARE GREATER.	
WAVE HEIGHT CREST TO TROUGH IN FEET	1, 2, 4 WHITE CAPS FORM, 6, 8, 10, 15, 20, 25, 30, 40, 50, 60
SEA STATE AND DESCRIPTION (DOUGLAS)	1 SMOOTH; 2 SLIGHT; 3 MODER-ATE; 4 ROUGH; 5 VERY ROUGH; 6 HIGH; 7 VERY HIGH; 8 PRECIPITOUS

The height of waves is arbitrarily chosen as the height of the highest 1/3 of the waves.

From Undersea Technology (now Sea Technology) May, 1964, p. 37.

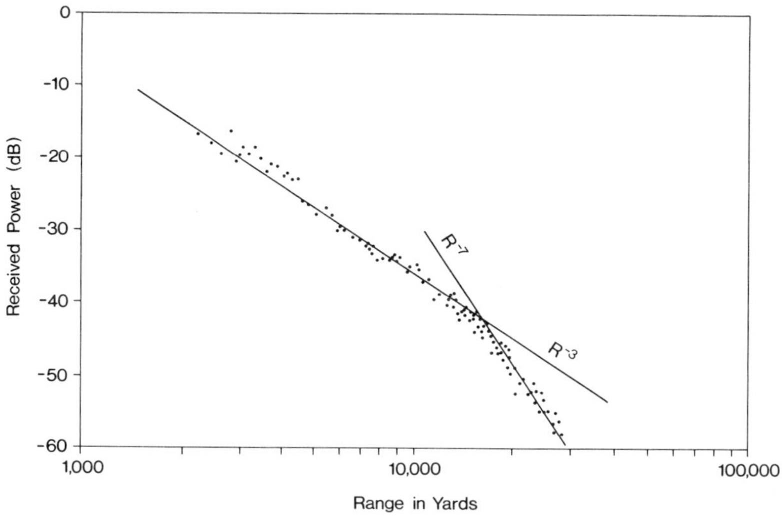

Fig. 6.19 Received sea clutter power versus range for 3.2 cm radar at 1,000 feet. Lines represent $1/R^3$ and $1/R^7$ variations (Katzin, 1957, On the Mechanisms of Radar Sea Clutter, Proceedings of the IRE, Vol. 45, No. 1, p. 45. Copyright © 1957 IEEE).

capillary waves act as Bragg scatterers where the Bragg resonance is achieved for:

$$\Lambda \cos \emptyset = \frac{n\lambda}{2} \qquad (6.12)$$

where Λ = ocean wavelength

λ = radar wavelength

$\emptyset$ = grazing angle

n = order of Bragg resonance.

It is usually assumed that only first-order Bragg scattering is significant. For grazing angles of 3° or less associated with surface-based radars, the Bragg resonant ocean wavelengths are about 1.5 cm, which correspondsto the wavelengths of the capillary waves.

The radar backscatter from the sea is usually specified as σ_0, which is the normalized radar cross-section and gives the radar cross-section per unit physical area of sea. The value of σ_0 is usually given in units of decibels with respect to one square metre; thus a σ_0 value of -10 dB would represent a radar cross-section of 0.1 m^2 per one square metre of physical sea area.

Experimental sea clutter measurements have been collected over many years by many experimenters. Nathanson (1969) has compiled a large number of the experimental results in tables of σ_0 for various sea states, frequencies, and grazing angles. The tables for the grazing angles of interest in this application are shown in Tables 6.4a and 6.4b. It must be emphasized that these are average values collected under various experimental conditions for a mix of upwind, downwind, and crosswind conditions.

Some of the trends in this data are more easily recognizable in the plots from Blake (1980) for sea state 3 as shown in Figure 6.20. Some general observations that can be made from the tables and figures are:

a) At low grazing angles (particularly below 20°), σ_0 increases with increasing frequency.
b) Below 15 or 20°, σ_0 increases with increasing grazing angle.
c) The vertically polarized σ_0 exceeds the horizontally polarized σ_0, but the difference between them decreases as the sea state, grazing angle, and frequency increase.
d) σ_0 increases rapidly as the sea state increases from low values (by as much as 10 dB/sea state) but reduces to a small change at higher sea state.

Sittrop (1977), amongst others, has shown the close relationship between sea clutter and wind speed. As it is accepted that a considerable portion of the sea

Table 6.4a Sea clutter backscatter for 0.1° and 0.3° grazing angles.

Normalized Mean Sea Backscatter Coefficient σ_0 for Grazing Angle of 0.1°

Sea state	Pol.	Reflection coefficient, dB, below 1 m^2/m^2 at indicated carrier frequency, GHz						
		UHF 0.5	L 1.25	S 3.0	C 5.6	X 9.3	K_u 17	K_a 35
0	*V*							
	H			90*	87*			
1	*V*					65*		
	H			80	75*	71*		
2	*V*	90*	87*	72*	64	56		
	H	95*	90*	75*	67*	61*		
3	*V*				56*	51		
	H	90*	82*	68	60*	53*		
4	*V*				53*	48		
	H			58	55	48		
5	*V*					44		
	H		65*	53	48	42*		

*5-dB error not unlikely. Monostatic radar, 0.5- to 10-μsec pulse.

Normalized Mean Sea Backscatter Coefficient σ_0 for Grazing Angle of 0.3°

Sea state	Pol	Reflection coefficient, dB, below 1 m^2/m^2 at indicated carrier frequency, GHz						
		UHF 0.5	L 1.25	S 3.0	C 5.6	X 9.3	K_u 17	K_a 35
0	*V*							
	H			83*	79	74*		
1	*V*			62*	60	58		
	H			74	71	66*		
2	*V*	80*		59*	55	52		
	H			66	60	56*		
3	*V*			55*	48	45		
	H			58*	50	46		
4	*V*			54*		43		
	H			50*		42	39*	
5	*V*	75*		50*		39		
	H			44	41	39	39*	

*5-dB error not unlikely. Monostatic radar, 0.5- to 10-μsec pulse.

(Nathanson, 1969, Radar Design Principles, McGraw-Hill. Reproduced by permission.)

Table 6.4b Sea clutter backscatter for 1.0° and 3.0° grazing angles.

Normalized Mean Sea Backscatter Coefficient σ_0 for Grazing Angle of 1.0°

Sea state	Pol.	Reflection coefficient, dB, below 1 m^2/m^2 at indicated carrier frequency, GHz						
		UHF 0.5	L 1.25	S 3.0	C 5.6	X 9.3	K_u 17	K_a 35
0	*V*		68*			60*		
	H	86*	80*	73	70	66		
1	*V*	70*	65*	56	53	50	47*	
	H	84*	73*	65	56	51	45	40*
2	*V*	63*	58*	53	47	44	42	38*
	H	82*	65*	55	48	46	41	38*
3	*V*	58*	54*	48	43	39	37	34
	H	76*	60*	48	43	40	37	36
4	*V*	55*	45	42	39	37	34	32
	H		52*	42	39	36	34	
5	*V*		43	38*	35	33	32	31
	H	65*	50*	42	35	33	31*	

*5-dB error not unlikely. Monostatic radar, 0.5- to 10-μsec pulse.

Normalized Mean Sea Backscatter Coefficient σ_0 for Grazing Angle of 3.0°

Sea state	Pol.	Reflection coefficient, dB, below 1 m^2/m^2 at indicated carrier frequency, GHz						
		UHF 0.5	L 1.25	S 3.0	C 5.6	X 9.3	K_u 17	K_a 35
0	*V*				60*	56*	52*	48*
	H	75*	72*	68*	63*	58*		53
1	*V*	60*	53*	52	49	45	43	41
	H	70*	62*	59	54	48	45*	43*
2	*V*	55*	53	49	45	41	39	37
	H	66*	59	53	48	42	41	40
3	*V*	43*	43	43	40	38	36	34
	H	61*	55*	46	42	39	37	37
4	*V*	38*	38	38	36	35	33	31
	H	54*	48*	41	38	35	34*	34
5	*V*		38	35	33	31	31*	30*
	H	53*	46	37	34	32	30*	

*5-dB error not unlikely. Monostatic radar, 0.5- to 10-μsec pulse.

(Nathanson, 1969, Radar Design Principles, McGraw-Hill. Reproduced by permission.)

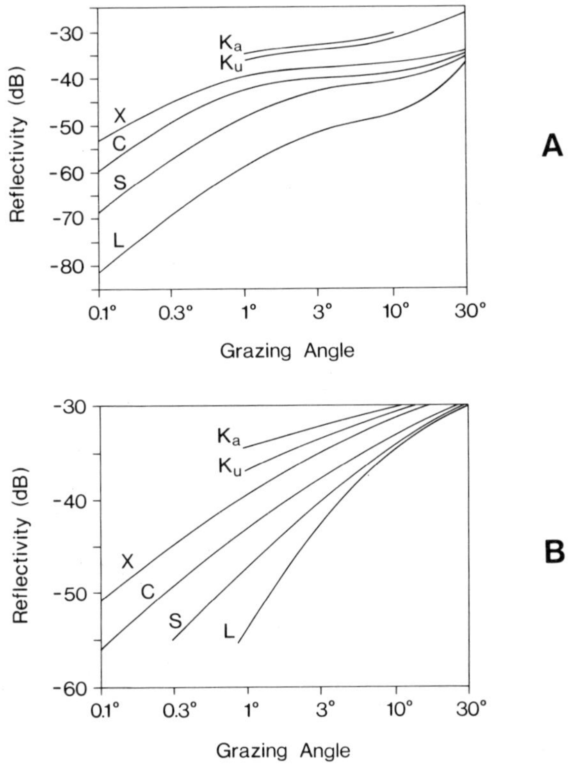

Fig. 6.20 Sea reflectivity (σ_o) as a function of grazing angle: for sea state 3 at L-, S-, C-, X-, K_u-, and K_a-bands for A) horizontal polarization and B) vertical polarization (Blake, 1980).

clutter return is from capillary waves and as the capillary waves are instantaneously excited by the wind, then it is not surprising that the sea clutter returns should be more strongly coupled to the wind velocity than the sea state. Capillary waves begin to form at a wind speed of about 3 m/s (6 knots) and increase in size until they reach a maximum size (on the order of centimetres) at about 24 m/s. As the wind speed abates below 3 m/s, the capillary waves die out quickly and the sea clutter returns reduce accordingly, even though the longer-period gravity waves will continue unchanged for some period of time. This point was emphasized by Glaser and Havig (1981) who noted that "smooth swells of even large (10 ft) amplitude are known to cause only small backscattering if no wind is 'roughening them.'" It would, therefore, appear that wind speed categories are a more appropriate method for sea clutter definition than sea state, particularly for abating winds.

Experimental Results

During the 1984 open-water radar experiment, some sea clutter data were collected at X-band. However, the relatively short fetch of the sea at the site and the low wind speeds experienced during the program produced only low sea states.

With wind speeds of 12 knots and wave height (peak-to-trough) of 80 cm (which might approximate sea state 2) at a grazing angle of 1°, the sea clutter HH σ_0 was about -49 dB. (A two-letter pair is used to denote polarization: the first letter for transmission and the second for reception.) This value of HH σ_0 is 3 dB below the values tabulated in Table 6.4b. The difference can be explained partially by the fact that this sea clutter data was collected cross-wind. Also, Table 6.4b is for fully developed seas, which were not present at this coastal site. Of equal interest is the fact that the average HH/HV ratio was only 7 dB. Thus the sea clutter can be seen to be a strong depolarizer.

A second set of sea clutter data (also HH-HV) but with wave height of 30 cm and wind speeds of 9 knots (sea state 2) yielded HH σ_0 for sea clutter at 1° of -73 dB, which is about 27 dB below the values of Table 6.4b. It is particularly interesting to note that the cross-polarized σ_0 for this data set is slightly higher

(-72 dB) than the like-polarized σ_0. A third data set with 18-cm wave heights and wind speeds of 7 knots (sea state 1) and 1° grazing angle yielded an HH σ_0 of -76 dB while the HV σ_0 was as high as -70 dB, or 6 dB higher.

Changing polarization, the VV σ_0 for sea clutter was significantly higher than for HH operation. For the third data set the average VV σ_0 value at 1° grazing angle was -55.7 dB, about 18 dB higher than the corresponding HH σ_0. The cross-polarized returns yielded VH σ_0 values of about -75 dB, very similar to the HV σ_0 value. The VV σ_0 for sea clutter is about 6 dB below the value in Table 6.4b.

Histograms were accumulated from all the sea clutter data. The average σ_0 for sea clutter for like polarization was -66.1 dB with a spread of values of 35 dB. The average cross-polarized σ_0 value was -74.0 dB with a spread of 30 dB.

Figure 6.21 shows a plot of power return from sea clutter for H and V transmission with the corresponding cross-polarized returns. These data were collected at sea state 2 with wind speeds of about 9 knots. The plot represents one sweep of the radar out in range from 750 to 2,000 m. The data were acquired for a long pulse (1 μs) transmission digitized at 5 MHz. The like-polarized returns are in the solid trace while the cross-polarized returns are in the dotted trace. It is apparent that the vertical-transmit/vertical-receive (VV) returns are 15 to 20 dB stronger than the horizontal-transmit/horizontal-receive (HH) returns. Looking at the top trace, the cross-polarized (HV) return is considerably stronger than the like-polarized (HH) return.

These data reflect a significant difference between horizontally- and vertically-polarized returns, which is not reflected in Table 6.4.

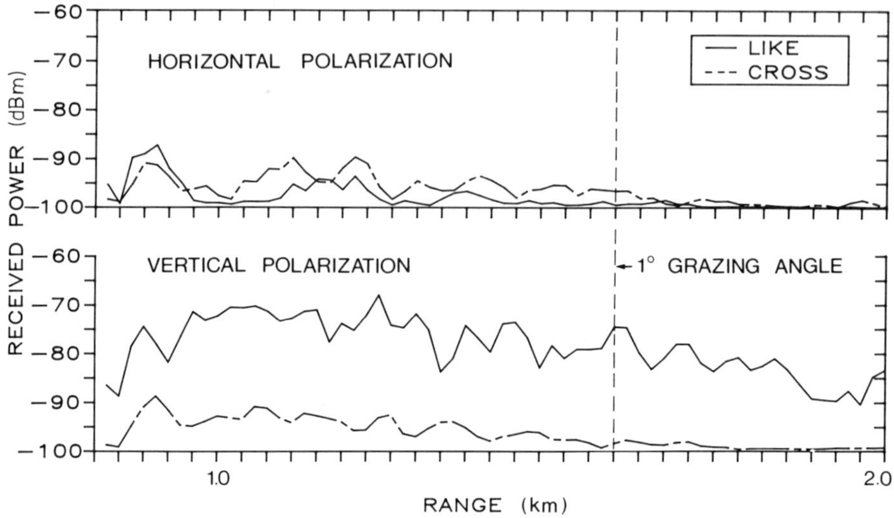

Fig. 6.21 Sea clutter returns for horizontal and vertical transmit. Solid traces are like-polarized returns, dotted trace is cross-polarized return (1984).

Sea Clutter Statistics

Once a value is selected for σ_0, it is then necessary to investigate the statistics of the sea clutter returns in order to apply them to determining probability of detection. If the number of scatterers contributing to the sea return is large, and if no single (or limited number of) scatterer returns a large part of the signal, then the returns can be considered as being Gaussian distributed. According to Blake (1980), the distribution can be considered as being Gaussian for as few as 10 principal scatterers. This Gaussian probability distribution is converted, upon passing through a detector, into a Rayleigh distribution, for which the probability density function is:

$$p(\upsilon) = \frac{\upsilon}{\sigma^2} e^{-\upsilon^2/(2\sigma^2)} \qquad (6.13)$$

where σ is the standard deviation of the signal (envelope) amplitude, υ. The Rayleigh distribution has been used historically to describe the statistics of sea clutter returns. For low range-resolution (long pulses), moderate antenna beamwidths, and low to moderate sea states, the Rayleigh distribution describes the statistics of sea clutter returns reasonably well. However, as these conditions are violated, the statistics become increasingly non-Rayleigh. In particular, as the resolution is increased or the sea state increases, there is a greater probability of large values of sea clutter than the Rayleigh distribution predicts. This results in longer "tails" on the distribution. An alternative probability distribution function that has been applied to sea clutter is the log-normal distribution. However, the log-normal distribution tends, in fact, to overestimate the occurrence of large values. The Weibull distribution has recently been suggested (Fay *et al.*,1977) as an appropriate distribution since it is intermediate between the Rayleigh and the log-normal. The probability density function for Weibull-distributed clutter amplitude is:

$$p(\upsilon) = \alpha \ln 2\left(\frac{\upsilon}{\upsilon_m}\right)^{\alpha-1} \exp\left[-\ln 2\left(\frac{\upsilon}{\upsilon_m}\right)^{\alpha}\right] \qquad (6.14)$$

where α relates to the skewness of the distribution and υ_m is the median value. The Rayleigh distribution is actually a member of the Weibull distribution and occurs when $\alpha = 2$.

Observations of radar A-scope displays have shown a "spikiness" to sea clutter returns even when the resolution is low and Gaussian distribution is assumed. In general, the returns are noise-like, although spikes are occasionally target-like in nature. The spikiness is more pronounced for horizontal polarization and increases with decreasing area of illumination.

The effect of decreasing pulse length on the distribution of the returns can be seen in Figure 6.22. The ratio of the standard deviation to the mean of the returns is plotted versus pulse length. For the Rayleigh distribution this ratio is one. As the pulse length is decreased, the standard deviation-to-mean ratio increases, indicating that the variability of signal return also is increasing. It is apparent that this is particularly true for horizontal polarization.

Investigations of the mechanisms causing sea spikes by Lewis and Olin (1980) have shown that large backscatter amplitudes are associated with the development and decay of whitecaps. They show that sea spikes usually occur coincidentally with the formation of whitecaps, although for vertical polarization the spike can occur just before formation of the whitecaps. Other experiments have reported strong spikes from unbroken sea waves. Ewell

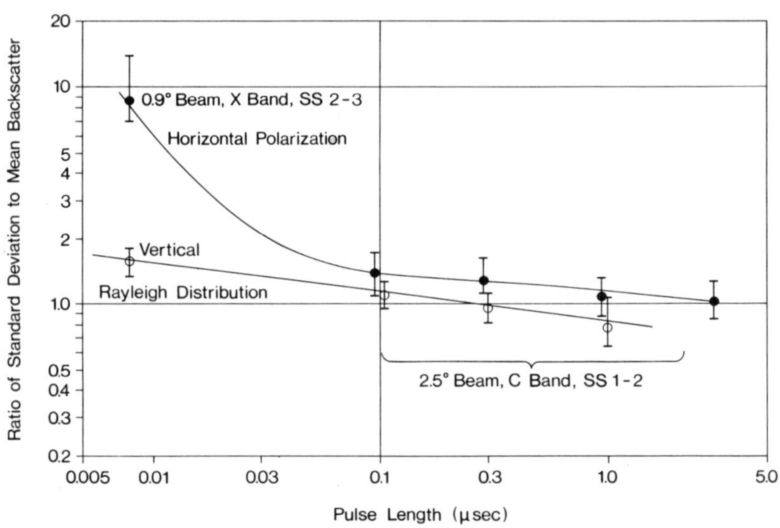

Fig. 6.22 Distribution of short-pulse sea backscatter, σ_0, standard deviation-to-mean ratio (Nathanson, 1969, Radar Design Principles, McGraw-Hill. Reproduced by permission).

et al. (1984) reported tracking of sea spikes with a high-resolution radar system for up to 18 s, although other experimenters have reported spike durations of up to 40 s.

For low grazing angles the largest backscatter occurs when the sea is observed in an upwind direction. Blake (1980) notes that the upwind/downwind σ_0 ratio may be as great as 10 for low grazing angles. Glaser and Havig (1981) have recorded sea clutter returns for all azimuths (as shown in Figure 6.23 for a wind speed of 22 knots, for which the upwind/downwind difference is obvious). Glaser and Havig noted that for a windspeed of 13 knots, the upwind clutter had decreased by 8 dB, while the downwind clutter had decreased by 30 dB. They commented that "downwind clutter is much more sensitive to wind speed changes than upwind." Long (1983) noted that upwind and downwind clutter had longer tails in the distribution than cross-wind clutter.

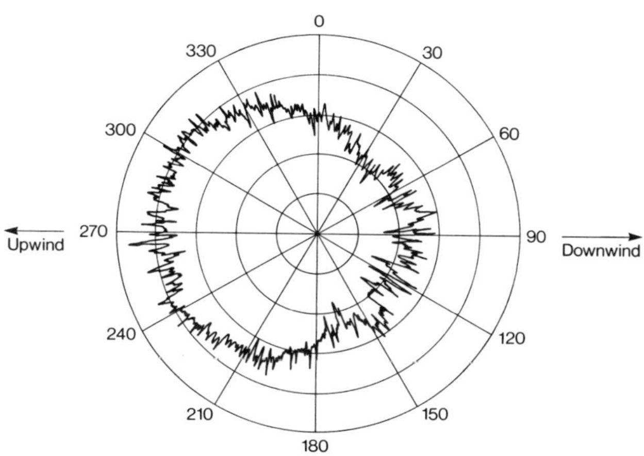

Fig. 6.23 Sea clutter radar cross-section (σ_0) versus azimuth for 22-knot wind (Glaser and Havig, 1981, Radar Backscattering from Ocean Waves at Low Grazing Angles, NACON Record, Vol. 1. Copyright © 1981 IEEE).

The rapid change in the sea clutter amplitude, with the exception of long-duration sea spikes, results in clutter returns decorrelating in about 10 ms. This means that samples taken more than 10 ms apart can be considered to be uncorrelated. As will be shown later, this timing becomes important when attempting to improve target detection in clutter by integration.

From a spatial perspective, Nathanson (1969) showed that samples about a pulse width apart in range are uncorrelated, while Maaloe (1982) suggested that the decorrelation distance is one to two resolution cells. In azimuth, Maaloe (1982) noted that the decorrelation distance varied from six beamwidths close to the radar to one beamwidth at far range, but the physical distance (in metres) was constant.

6.5 SUMMARY

This chapter reviewed several environmental factors affecting surface-based radar performance. The chapter began with a description of multipath interference and presented experimental results which confirmed the predicted theoretical behaviour.

The chapter continued with a review of atmospheric attenuation and anomalous propagation (ducting; subrefraction) due to refractive index changes. This section included an analysis of meteorological data from the east coast of Canada which indicated that sub-normal radar conditions occur with a high frequency in this area. The chapter concluded with a discussion of sea clutter and presentation of some limited sea clutter data collected in the Arctic.

CHAPTER 7
Radar Frequency Effects

When considering the choice of frequency, or frequencies, for radar systems for ice detection and classification, there are several practical considerations and at least three, and possibly four, phenomenological considerations which need to be taken into account. The first part of the chapter deals with these considerations, then the chapter concludes with a discussion of experimental results.

7.1 PRACTICAL CONSIDERATIONS

One of the practical considerations pertains to the availability and cost of radar systems at frequencies above X-band (10 GHz). The cost of radar hardware above 10 GHz is expected to increase exponentially. In addition to the cost factor, there is a further difficulty in obtaining high-power transmit devices as the frequency increases. This problem can be partially offset by the increasing gain of the antenna for the same physical aperture as the frequency increases. An additional technical problem is the increase of losses in the waveguide as the frequency increases. This is particularly a problem in experimental programs where installations require the use of high-loss flexible waveguide.

There are, however, some accrued advantages as the frequency increases. As shown in Chapter 6, the effects of multipath interference are reduced as the frequency increases, because the height of the first maximum (Figure 6.10) moves closer to the surface. Also, as the frequency increases, the azimuthal beamwidth decreases for the same physical antenna aperture, which not only improves azimuthal resolution, but also reduces the effect of sea clutter, since the radar cell size is reduced.

7.2 PHENOMENOLOGICAL CONSIDERATIONS

The phenomenological considerations of interest are: surface (Rayleigh) roughness as a function of frequency, skin depth, atmospheric attenuation, and the possibility of resonance between the radar signal and ice features.

Rayleigh Roughness

The energy incident on a surface will be reflected in one of two distinct fashions. If the surface is "smooth," the energy will be reflected specularly in a single direction away from the radar, but if the surface is "rough," the energy will be diffused and scattered in many directions, as shown pictorially in Figure 7.1. This effect can be used to define rough and smooth surfaces. A surface that reflects the incident energy specularly will be considered smooth, whereas a surface that scatters the energy in various directions will be considered rough. From the perspective of the incident energy, the smoothness or roughness of a surface is dependent on three variables: the wavelength of the incident energy, the angle of incidence, and the physical relief of the surface.

The Rayleigh roughness criterion relates these variables and provides a mathematical determination of

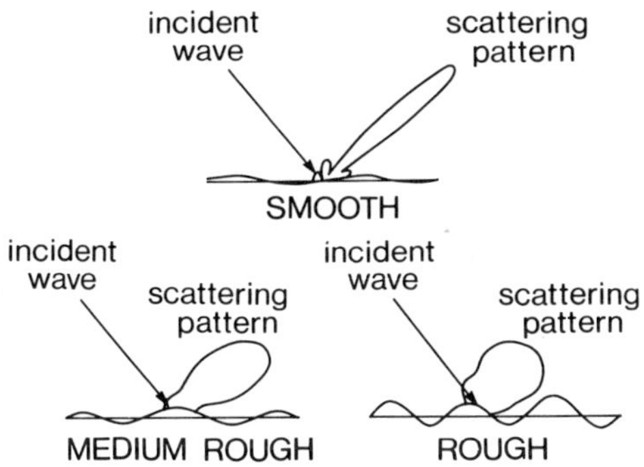

Fig. 7.1 Surface scattering patterns as a function of surface roughness. Reprinted by permission of Artech House, Inc., from F.T. Ulaby _et_ _al_., Radar Remote Sensing and Surface Scattering and Emission Theory, Vol. 2, p. 819. Copyright © 1982 by Addison-Wesley Publishing Co.

surface roughness. If the surface is considered as shown in Figure 7.2, the path difference between a ray striking the top of the surface and another striking the bottom can be defined as:

$$\Delta r = 2h \sin\psi \tag{7.1}$$

where h is the height of the surface feature, and ψ is

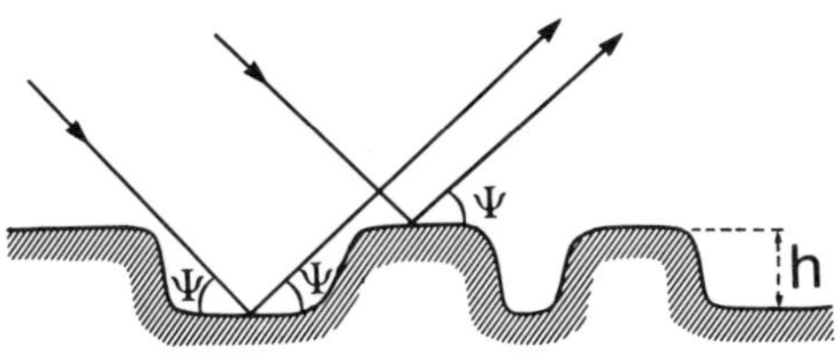

Fig. 7.2 Derivation of the Rayleigh roughness criterion (Reprinted with permission from Beckmann and Spizzichino, The Scattering of Electromagnetic Waves from Rough Surfaces, 1963, Pergamon Press).

the grazing angle. This path difference can be converted to a phase difference between the two rays as

$$\Delta\phi = \frac{2\pi}{\lambda}\Delta r = \frac{4\pi h}{\lambda}\sin\psi \qquad (7.2)$$

If the surface is smooth, the two rays will be in phase. However, as the phase difference increases, the two rays will begin to interfere until at $\Delta\phi = \pi$ they will be out of phase and will cancel each other, and no energy will be reflected in the forward direction. By the law of conservation of energy, however, the energy has not been lost; consequently, it must have been scattered in other directions. It can, therefore, be said that on a surface where the phase difference between the ray striking the top of the surface and the one striking the bottom is π, the surface is definitely rough. Conversely, where the difference is zero, the surface is smooth.

A division between rough and smooth can be established arbitrarily by choosing a point halfway between the two points, i.e., $\pi/2$ which then yields the normal Rayleigh criterion where a surface is considered smooth for

$$h < \frac{\lambda}{8 \sin \psi} \qquad (7.3)$$

Table 7.1 shows the relationship between the grazing angle, the frequency (or wavelength), and the surface height (in centimetres) for a surface which is considered rough. Note that for a 1° grazing angle, at S-band the surface is not rough until the surface features vary by 71 cm, but at K_a-band the surface only need vary by 6 cm to be considered rough. The practical implications of this roughness criterion will be shown later in this chapter.

Table 7.1 Rayleigh roughness criterion for radars of different wavelength.

Grazing angle (deg)	Roughness (cm) S-band 10.0 cm λ	X-band 3.0 cm λ	K_u-band 1.9 cm λ	K_a-band 0.86 cm λ
5.0	14.0	4.3	2.7	1.2
2.0	36.0	11.0	7.0	3.1
1.0	71.0	21.0	14.0	6.0
0.75	95.0	29.0	18.0	7.0
0.5	143.0	43.0	27.0	12.0
0.25	286.0	86.0	54.0	25.0

Skin Depth Versus Frequency

As shown in Chapter 3, the skin depth is the depth in the material at which the field strength of the incident energy is down to 1/e. This depth is a function of the dielectric properties of the material.

Table 7.2 shows the skin depths (in centimetres) for S- through K_a-band for various values of loss tangent, and ε_r' of 3.12, which was shown to be a typical value in Chapter 3. Typical ranges of loss tangent for first-year and multiyear ice are indicated. It can be seen that at S-band with a loss tangent of 0.01 (for multiyear ice) the skin depth is 180 cm, while at K_a-band the skin depth is 15.5 cm.

Atmospheric Attenuation

As discussed in Chapter 6, the presence of atmospheric gases and moisture can have a significant effect on the propagation of a radio wave through the atmosphere. For fog and rain, this effect is frequency dependent. As Table 7.3 (Saxton and Hopkins, 1951) shows, the attenuation caused by fog increases substantially as the radar

Table 7.2 Skin depth as a function of radar frequency and loss tangent.

Loss tangent (tan δ)	Ice type	Depth (cm) S-band 3 GHz	X-band 10 GHz	K_u-band 16 GHz	K_a-band 35 GHz
0.001	Glacier	1801.0	541.0	342.0	155.0
0.005	Glacier	360.0	108.0	68.5	31.0
0.010	Multiyear	180.0	54.1	34.2	15.5
0.020	Multiyear	90.1	27.0	17.1	7.7
0.040	Multiyear / First-year	45.1	13.5	8.6	3.9
0.060	Multiyear / First-year	30.0	9.0	5.7	2.6
0.080	First-year	22.5	6.8	4.3	1.9
0.100	First-year	18.0	5.4	3.4	1.6
0.150	First-year	12.0	3.6	2.3	1.0
0.200	First-year	9.0	2.7	1.7	0.8
0.300	First-year	6.0	1.8	1.1	0.5
0.400		4.5	1.4	0.9	0.4
0.500		3.6	1.1	0.7	0.3

Table 7.3 Attenuation caused by fog in polar regions*.

Visual range (m)	Attenuation (dB/km) When λ = 1.25 cm	When λ = 2.25 cm	When λ = 3.2 cm	When λ = 10 cm
30	1.2	0.40	0.20	0.02
60	0.45	0.15	0.08	0.008
90	0.25	0.08	0.04	0.004
150	0.12	0.04	0.02	0.002
225	0.07	0.02	0.01	0.001
300	0.045	0.015	0.007	0.001

* Temperature about 0°C.

(From Saxton and Hopkins, 1951)

frequency increases. This effect was also shown graphically in Figure 6.12. The attenuation by atmospheric gases as a function of frequency is shown in Table 7.4 (Saxton and Hopkins. 1951). Again, the attenuation increases as the frequency increases.

Table 7.4 Attenuation caused by atmospheric gases.

	Attenuation (dB/km)			
	Polar region		Temperate region	
Wavelength (cm)	Typical	Maximum	Typical	Maximum
1.25	0.09	0.15	0.2	0.35
2.25	0.015	0.02	0.025	0.035
3.20	0.01	0.01	0.015	0.02
10.00	0.007	0.007	0.007	0.007

(From Saxton and Hopkins, 1951)

Finally, the attenuation resulting from rainfall, as shown in Figure 7.3, increases as the wavelength decreases (frequency increases). For a heavy rainfall of 16 mm/hr, the attenuation per kilometre varies from less than 10^{-2} dB at S-band, to about 6 dB at K_a-band.

Resonance

The last effect to be considered is that of resonance. As the wavelength of the radar signal begins to approach the size of physical features in the ice, there is an anticipated increase in the amount of energy returned. The ice features that could contribute to this effect include the general micro-surface roughness of the ice, as well as the inhomogeneities in the ice (air bubbles, brine drainage channels, etc.). The resonant effect from the surface is, in fact, directly related to the Rayleigh roughness developed earlier. It is anticipated that the

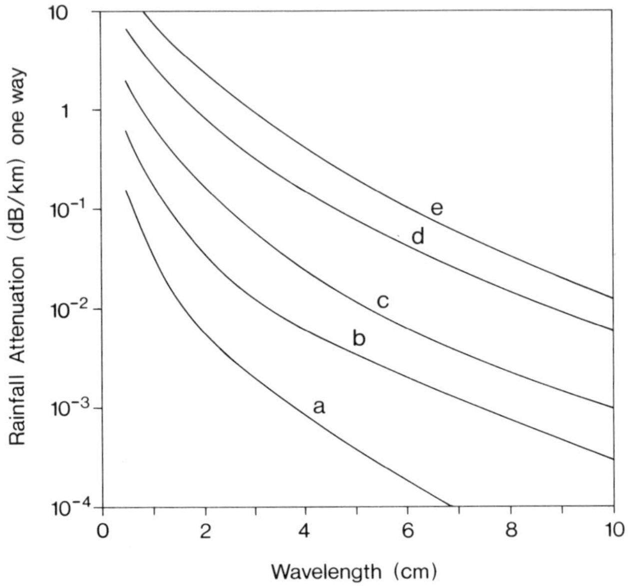

Fig. 7.3 One-way attenuation in rain at 18°C: for a) drizzle - 0.25 mm/hr; b) light rain - 1 mm/hr; c) moderate rain - 4 mm/hr; d) heavy rain - 16 mm/hr; e) excessive rain - 40 mm/hr (from Skolnik, 1980, Introduction to Radar Systems, McGraw-Hill).

wide range of sizes of surface features will cause a resonance, or surface features will appear rough, at all microwave frequencies. However, the resonance with internal inhomogeneities in the ice will occur only over a limited range of frequencies.

7.3 EXPERIMENTAL RESULTS

Analysis of the data collected during the 1983 experiment shows substantial differences in the radar returns from the ice surface as a function of frequency. Figure 7.4 shows the aerial mosaic for a section of ice corresponding to Standard Line 247 and the corresponding radar returns for several frequencies.

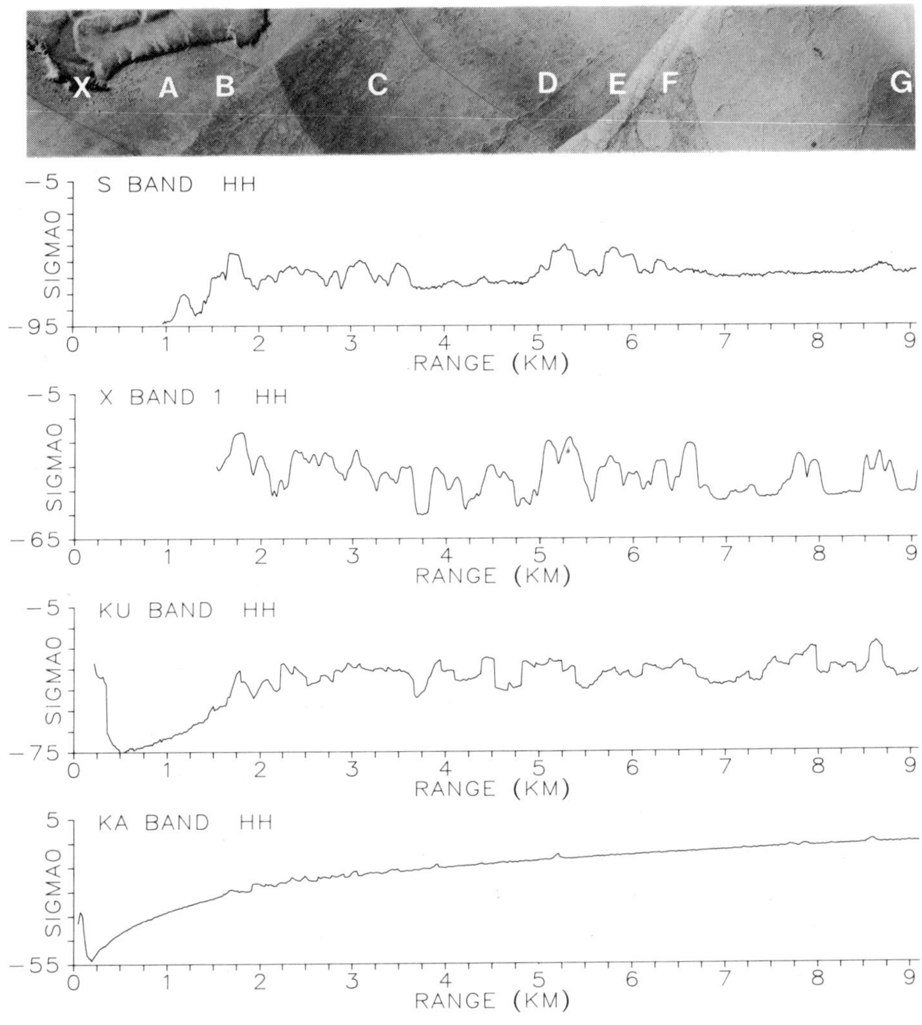

Fig. 7.4 Photomosaic and radar returns along Line 247 for S-, X-, K_u-, and K_a-bands. Radar is located at X on left of figure. Area from A to B is smooth first-year (FY) ice, B to C rough FY ice, C to D smooth FY ice, medium-sized pressure ridges at D and E, rough FY ice from E to F, and a large multiyear floe from F to G.

The transmitter is located at X on the left side of the image, while the range from the transmitter to the right side of the mosaic is about 9.0 km. The ice surface depicted consisted of relatively smooth, first-year ice in the area A to B, rougher first-year ice from B to C, smooth first-year ice from C to D, medium-sized first-year pressure ridges at D and E, rough first-year ice in the area E to F, and a large multiyear floe in the area F to G.

The normalized radar cross-sections, σ_0, for horizontal transmit/horizontal receive for S-, X-, K_u-, and K_a-band are shown below the mosaic. The X- and K_u-band plots are very 'busy,' with substantial returns from all along the line. From a review of the aerial mosaic and extensive surface verification, it was concluded that many of these strong returns were caused by targets that would not be significant from a navigational point of view; that is, they would not represent a threat to a ship or be a major impediment to navigation.

The S-band plot, on the other hand, is considerably less busy. The strong signal returns in the region B to C correspond to significant roughness in this area, visible on the aerial photograph and substantiated by surface investigation, while the low signal returns from C to D correspond to the smooth area on the mosaic. The twin ridges at D and E are well represented by strong peaks. The roughness immediately behind the ridges is also displayed, as well as the smoothness of the multiyear floe. A multiyear ridge in the centre of this floe is also evident.

There are two frequency effects that contribute to the generally lower returns at lower frequencies. First, the Rayleigh roughness criterion specifies that targets small with respect to the wavelength will not appear "rough" to the radar wave and thus will return little energy. As the wavelength of the radar signal decreases, targets will appear increasingly rough and will reflect more energy. Second, as the frequency increases the lower side of the first multipath interference lobe will move closer to the surface, putting more energy on the surface.

The K_a-band returns were more complex. As indicated in Chapter 5, the K_a-band radar suffered from low output power and high waveguide losses, resulting in weak returns. This is evident in Figure 7.4 and makes the interpretation of the K_a-band returns difficult. However, the returns, although at low levels, are significant. The targets associated with the returns in the area F to G were multiyear pressure ridges on the multiyear ice floe. The return at D was caused by a substantial first-year pressure ridge which contained a considerable number of multiyear fragments.

An analysis of the returns from a second standard line, Line 341, yielded similar results. From a navigational point of view, the targets of interest (Figure 7.5) were an iceberg at B and multiyear floes at A and C. Again it was evident that the signal returns became more cluttered from S-, to X-, to K_u-band because the surface appeared increasingly rough as the frequency increased, and as the frequency increased the first interference lobe moved closer to the ice. The K_a-band return was reduced significantly by the low power of the transmitter and the high waveguide losses.

The S-band returns reflected the general roughness of the ice. The S-band radar clearly detected the iceberg at B but missed the multiyear floe at A. The S-band radar detected the ice floe at C because of its wider beamwidth, while the other systems did not illuminate the floe. The X- and K_u-band systems gave good returns from the iceberg but did not show an ability to detect

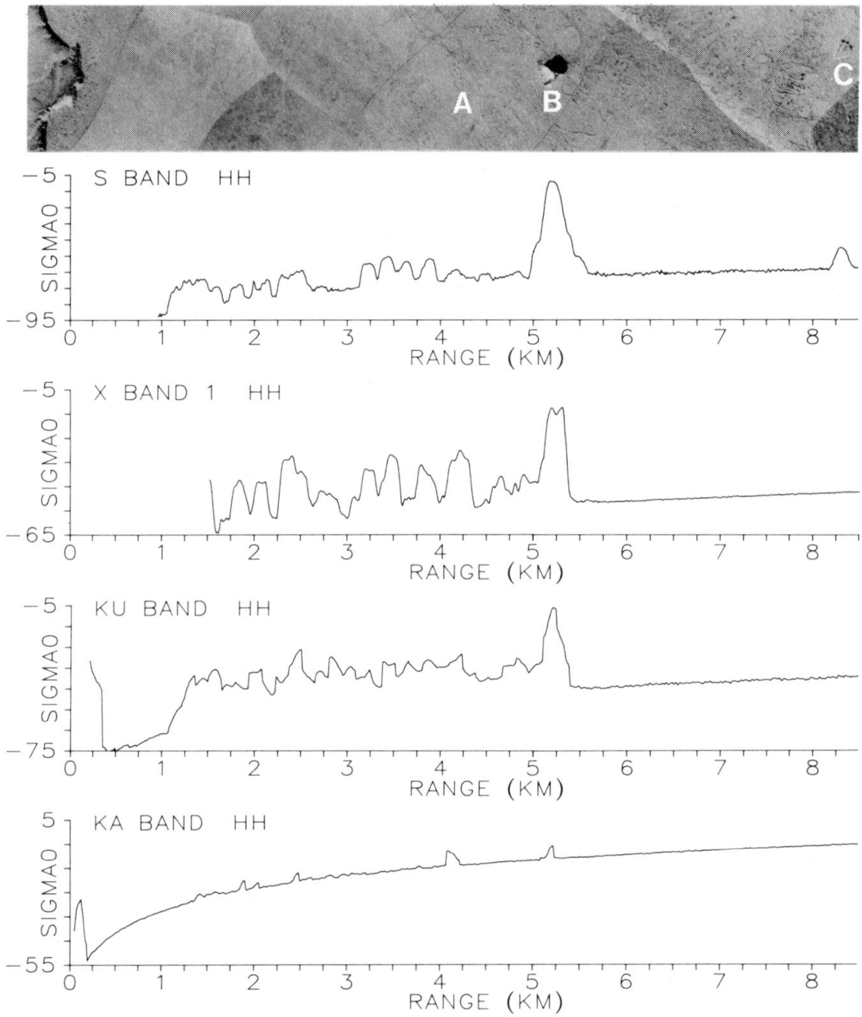

Fig. 7.5 Photomosaic and radar returns along Line 341 for S-, X-, K_u-, and K_a-bands. Targets of interest are multiyear floes at A and C and iceberg at B.

other targets. The K_a-band, although plagued by low-level returns, detected the iceberg at B and the multiyear floe at A.

The analysis of several other standard line data sets confirmed the findings discussed above. The S-band system gave returns that were strongly related to the target's vertical extent, thus displaying targets which would be a threat to navigation. However, the S-band system did not detect multiyear ice floes that were without significant surface relief. The X- and K_u-band returns showed responses to surface features that were much smaller in vertical extent (many of which would not have been a threat from a navigational point of view) than those detected by the S-band radar. The X- and K_u-band returns presented a cluttered response that was not easy to interpret. The K_a-band returns, on the other hand, although lower in level, correlated well with targets of interest, including icebergs, multiyear ice floes, and some first-year pressure ridges.

The normalized, like-polarized radar cross-section values (σ_0) for first-year, multiyear, and glacier ice are shown in detail in Table 7.5. As expected, the average σ_0 values for each ice type increased with frequency. Although the values themselves are interesting, it is more informative at this point to examine the change in σ_0 for each ice type as the frequency increases. These results are shown in Table 7.6, and although they will be examined in some detail in Chapter 9, it is worth noting here that the change in the σ_0 for glacier ice between S- and K_a-band, X- and K_a-band, and K_u- and K_a-band is considerably less than for first-year and multiyear ice. In addition, there are, in general, systematic differences in the change of σ_0 for multiyear and first-year ice in the three frequency pairs. With the exception of the change

Table 7.5 Radar cross-section (σ_0) of ice versus frequency.

Feature	Radar cross-section (dB)			
	S-band 3 GHz	X-band 10 GHz	K_u-band 16 GHz	K_a-band 35 GHz
Iceberg				
Line 292	-41	-20	-6	-13
Line 307 (1)	-21	-22	-11	-6
(2)	-54	-35	-36	-17
Line 341	-9	-13	-6	-6
Mean	-31	-22.5	-14.8	-10.5
Multiyear ice				
Line 247 (1)	-56	-30	-22	-4
(2)		-30	-25	-6
Line 277 (1)	-60	-26	-27	-12
(2)	-53	-33	-23	-5
Line 292	-56	-30	-26	-8
Line 341	-63	-30	-28	-8
Mean	-57.6	-29.8	-25.3	-7.2
First-year Ice				
Line 0	-61	-38	-30	-18
Line 247 (Ridge)	-45	-23	-32	-10
Line 277 (Ridge)	-39	-23	-30	-13
Mean	-48.3	-28.0	-30.7	-13.7
Island	-35	-22	-21	-5

Table 7.6 Change in radar cross-section of ice as frequency is changed.

Feature	Change in σ_0 (dB)		
	from S to K_a	from X to K_a	from K_u to K_a
First-year ice	35	14	16
Multiyear ice	50	22	18
Glacier ice	20	12	4

between K_u- and K_a-band (which may be affected by poor K_a-band returns), the change in the σ_0 for multi-year ice is more substantial than for first-year ice.

Analysis of the standard line data gives a quantitative measure of the effect of frequency on the detection and classification of ice features, and a visual examination of images for each of the frequencies supports the line data. Figure 7.6 shows corresponding segments of images from the 1983 experiment for each of S-, X-, K_u-, and K_a-band for horizontal polarization.

Surface verification of the area established that the targets of interest from a navigational perspective are the first-year pressure ridge at A; icebergs at B, C, D, and E; an island at F; and a large number of multiyear ice floes distributed throughout the ice field. Examination of the images shows that the S-band image best defines the first-year pressure ridge, as well as adequately displaying the four icebergs and the island. The smallest of the icebergs, D, however, is somewhat difficult to pick out from the other clutter. The S-band image also highlights several of the multiyear floes north of (above) the pressure ridge. The X- and K_u-band images show returns from a large number of targets, many of which are not significant navigational threats. These returns in many cases conceal returns from targets of interest. Neither the X- nor the K_u-band image clearly displays the first-year pressure ridge. In addition, the icebergs and island are more difficult to find, particularly in the X-band image. Also, the small iceberg at D is lost in both the X- and K_u-band images.

Finally, although the signal levels received with the K_a-band system were low, all the icebergs (including the small one at D) as well as the island at F, are identifiable on the K_a-band image. Although the returns from far range targets are low, the targets that are

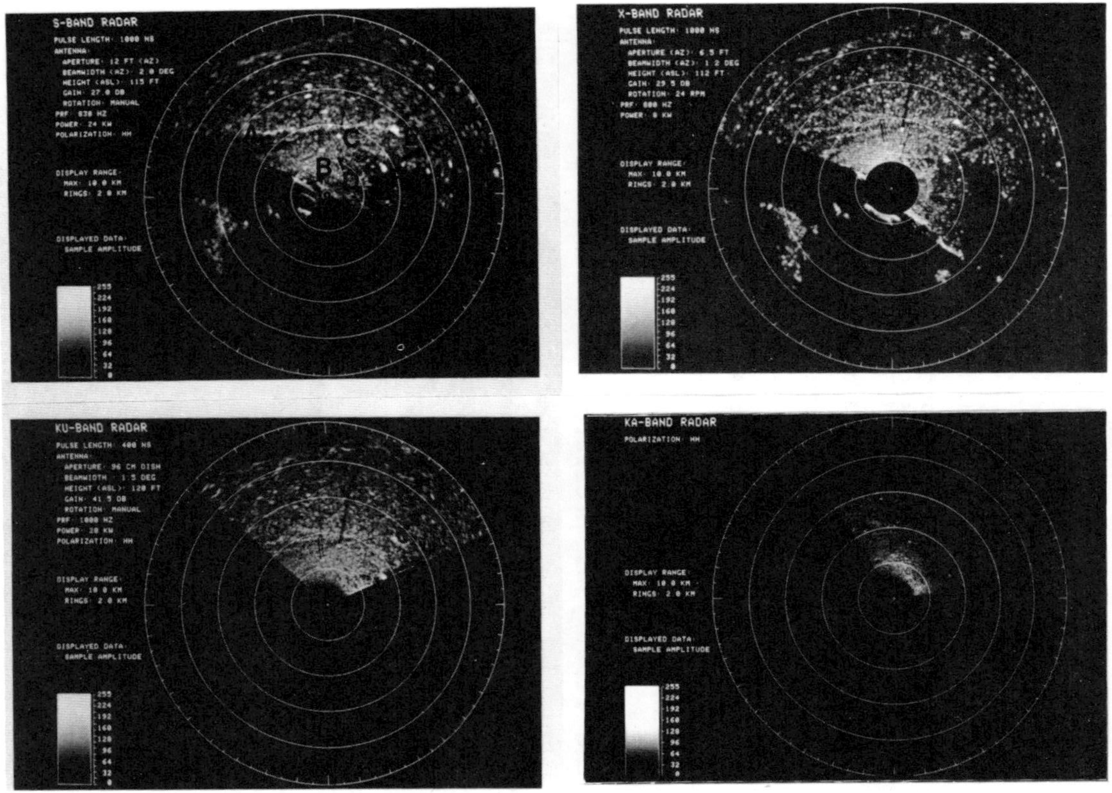

Fig. 7.6 Images of radar returns from ice using S-, X-, K_u-, and K_a-band. Targets of interest are first-year pressure ridge at A; icebergs at B, C, D, and E; and island at F.

displayed at G, H, and J correspond to known multiyear floes.

7.4 SUMMARY

This chapter addressed frequency effects as they influence the performance of surface-based radar systems operating in an ice environment. The chapter began by reviewing some practical considerations, particularly as they relate to higher frequency systems. The chapter continued by looking at four phenomenological effects: Rayleigh roughness, skin depth, atmospheric attenuation, and resonance.

The chapter concluded with analysis of horizontally polarized ice radar data at S-, X-, K_u- and K_a-bands. Values of the normalized radar cross-section for various ice types at each frequency were given.

Examination of radar images of the ice field at each of the frequencies showed that the S- and K_a-bands provide information about the ice not apparent in the other frequency images. The K_a-band in particular was able to enhance the detectability of the icebergs and some of the multiyear ice floes.

CHAPTER 8
Radar Polarization Effects

8.1 INTRODUCTION

There are five parameters of the electromagnetic wave returned from a target that can be used to extract information about the target. They are: frequency, amplitude, phase, Doppler, and polarization. For a non-coherent system (as employed in conventional marine radars) these parameters are reduced to frequency, amplitude, and polarization. Conventional marine radar systems operate at both 9.4 GHz (X-band) and 3 GHz (S-band), and use amplitude returns. However, all conventional marine radars have a single fixed polarization that is invariably horizontal.

Cloude (1983) defines polarization as "the spatially directive property of a wave." This definition can be expanded to state that the polarization is the orientation of the electric field vector at a point in space during one oscillation. Consequently, it is reasonable to assume that the polarization state of the return signal from a target might yield information about the target symmetry and orientation. Again, as noted by Cloude (1983), a target can be considered as a polarization transformer. Therefore, if the full polarization state of the returned signal were known, considerable information about the target could be obtained.

In general, there are three types of polarization normally considered. These are linear, circular, and elliptical. In a plane polarized wave, the polarization is specified by the orientation of the electric field components of the electromagnetic wave. Thus an electromagnetic wave with the electric field oriented horizontally is referred to as a horizontally polarized wave. The wave may also be vertically polarized or at any angle to the vertical or horizontal. A circularly polarized wave has an electric field of constant magnitude, rotating at the RF rate. Looking from behind the antenna, if the electric field were to rotate clockwise it would be classified as right-circular polarized (RCP), while if it rotates counter-clockwise it would be left-circular polarized. The more general, and in fact all-encompassing polarization state, is the elliptically polarized wave. As shown in Figure 8.1, the electric field intensity varies as it rotates and traces out an ellipse. The ellipse can be completely defined by the orientation, θ, and the axial ratio ($\tan \tau$). By convention τ is specified as ±45°, where positive τ represents

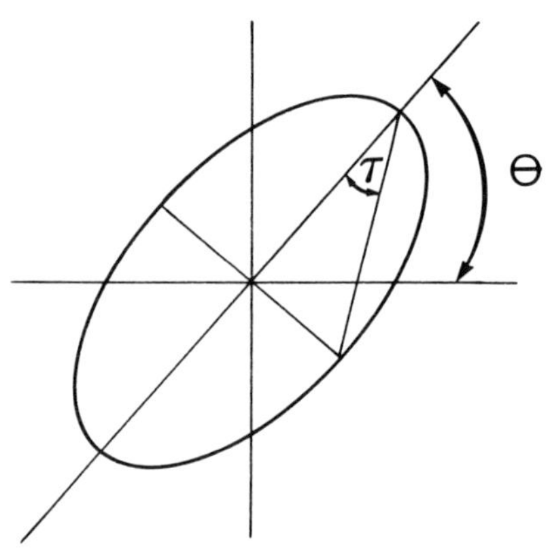

Fig. 8.1 Polarization ellipse, where θ is the orientation and $\tan \tau$ is the axial ratio.

left-hand polarization and negative τ represents right-hand polarization. As shown in Figure 8.2, if the axial ratio is zero, a linear polarized wave results, and with an orientation of 90° the wave is vertically polarized; if the axial ratio is one, a circularly polarized wave results.

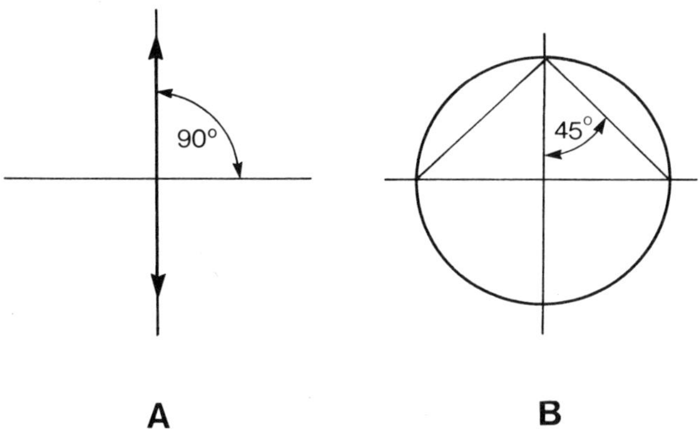

Fig. 8.2 Special polarization ellipses: A) for tan τ equals 0 and θ equals 90° resulting in a vertically polarized wave; B) for tan τ = 1 resulting in a circularly polarized wave.

In general, when a linear, circular, or elliptically polarized wave strikes a target, the reflected wave will be elliptically polarized. In addition, if the target depolarizes the transmitted wave, considerable energy will be at a polarization different from the transmitted polarization. It is obvious, then, that a single polarized antenna will not receive all the returned energy. In fact, an antenna system capable of receiving two orthogonally-polarized signals is required to capture all the returned energy.

In specifying the returned signals in a dual-polarized system, the convention has been to specify the transmit and receive polarizations as subscripts to the parameter being measured. The first subscript specifies the transmit polarization and the second, the receive polarization. Thus E_{HV} would specify horizontal transmit-vertical receive.

Scattering Matrix

If a coherent transmitter is used, in addition to the amplitude information, phase information can also be collected for each of two orthogonal polarizations. If the two polarizations are horizontal (H) and vertical (V), a coherent scattering matrix can be written to represent the backscattering properties of the target:

$$M = \begin{bmatrix} (S_{HH}) \exp(j\theta_{HH}) & (S_{HV}) \exp(j\theta_{HV}) \\ (S_{VH}) \exp(j\theta_{VH}) & (S_{VV}) \exp(j\theta_{VV}) \end{bmatrix} \tag{8.1}$$

The scattering matrix is a generalization of the target's radar cross-section to include the target's dependence on polarization (Huynen, 1965). This matrix is dependent on the aspect angle of the target, the orientation of the target, and the frequency of transmission.

Using the denotation of Cloude (1983), the scattering matrix relates the transmitted field to the received field as:

$$\begin{bmatrix} E_H \\ E_V \end{bmatrix}_R = A \begin{bmatrix} |S_{HH}| \exp(j\theta_{HH}) & |S_{HV}| \exp(j\theta_{HV}) \\ |S_{VH}| \exp(j\theta_{VH}) & |S_{VV}| \exp(j\theta_{VV}) \end{bmatrix} \cdot \begin{bmatrix} E_H \\ E_V \end{bmatrix}_T \tag{8.2}$$

where the T and R signify transmit and receive, respectively. To determine the scattering matrix of Equation 8.2, it is necessary to measure four amplitudes and four absolute phases. However, a relative scattering matrix can be determined where, because of the reciprocity of S_{HV} and S_{VH} and measurement of relative phases, only three amplitudes and two relative phases need be measured. Huynen (1965) described measurement techniques where the scattering matrix can also be determined by the measurement of amplitude only, using six linear polarizations.

Using the scattering matrix, techniques have been developed which enable classification of targets by the properties of the scattering matrix, or improvements in the target-to-clutter performance by suppression of unwanted returns. However, these techniques all require sophisticated radar systems that are coherent or which employ multipolarization.

The radar equipment employed in the studies described in this book was non-coherent but in several cases was dual-linearly polarized. The remainder of this Chapter discusses the results achieved with these dual-polarized systems.

8.2 EXPERIMENTAL RESULTS

Early Indications

Analysis of the cross-polarized data at X-band collected in 1981 showed that there were distinct differences between the like- and cross-polarized replies. Figure 8.3 is an A-scope photograph of the returns from the Decca RM 1229 X-band radar, like-polarized (HH) on the lower trace and cross-polarized (HV) on the upper trace. The cross-polarized return was received with a parabolic dish rotated to receive vertical polarization and fed to the Furuno FRM 64W radar. The signals recorded were logarithmic raw video from each radar, with the radar gains adjusted to give non-saturating video of approximately equal amplitude for both traces. Consequently, the returns were uncalibrated and no comparison of absolute amplitudes can be made.

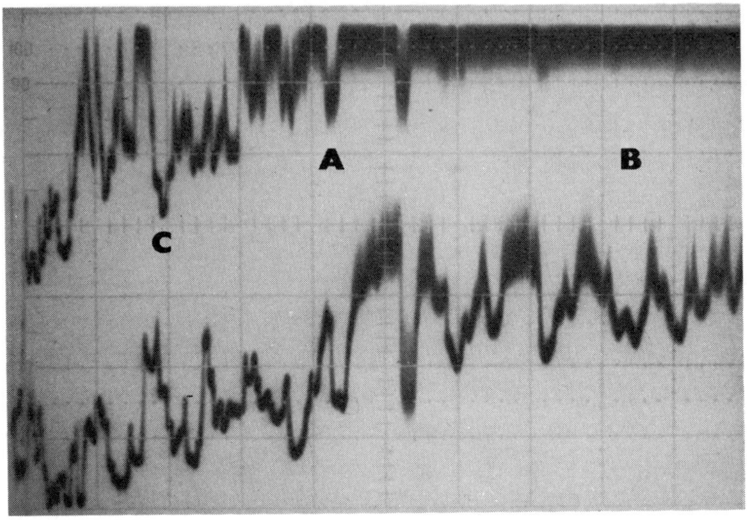

Fig. 8.3 A-scope trace of like- (bottom trace) and cross- (top trace) polarized radar returns at X-band along Line 1 from 1981 experiment.

To ensure that both antennas were illuminating the same ice surface, an artificial radar target (Luneburg lens) was deployed on the line and the antennas for both systems were aligned to give a maximum return from the target. The antennas were then locked into position for data collection. It is apparent by comparing the two traces that the like- and cross-polarized signals responded differently. For example, the target at A gave a strong response on the like- and cross-polarized channel, whereas the target at B, which gave a strong reply in the like-polarized channel, provided no response in the cross-polarized channel. Conversely, the target causing the return at C gave a much stronger return on the cross-polarized channel than on the like-polarized channel. Because of limited mobility on the ice it was not possible to confirm which ice features were causing the unusual returns on the cross-polarized channel.

These findings, however, prompted the major frequency/ polarization experiment conducted in 1983.

Full Ice Cover Results

Figure 8.4 shows a strip of the aerial photomosaic corresponding to Line 307 from 1983, with the three higher frequency (X-, K_u-, and K_a-band) returns for like- and cross-polarization shown below. Two targets of interest, which demonstrate the use of polarization to differentiate and detect ice features, are shown. The target at A was a large iceberg with a height of 45 m and width of 75 m, yielding a physical cross-section of about 3,400 m². The low signal levels to the right of the iceberg in the K_u- and K_a-band returns were the result of the radar shadow cast by the iceberg. The wider beamwidth of the X-band antenna illuminated targets which were not directly behind the iceberg. Target B to the left of the large iceberg was another small iceberg (300 m²) which would have been vitally important to detect as a hazard to shipping.

All three frequencies using like polarization detected the iceberg at A. However, by comparing the like- and cross-polarized returns it can be seen that, with the exception of the return from the iceberg, the cross-polarized signals are considerably lower than the like-polarized returns and this makes the iceberg more readily observable. The improvement in target-to-clutter ratio (where the clutter is considered to be all those signals from ice targets that are not of interest from a navigational point of view) means that the target was more easily differentiated from the other surrounding signals. In this situation the clutter was predominantly caused by returns from first-year ice. The improvement in target-to-clutter ratio is particularly apparent in the K_u-band returns where the improvement exceeds 10 dB.

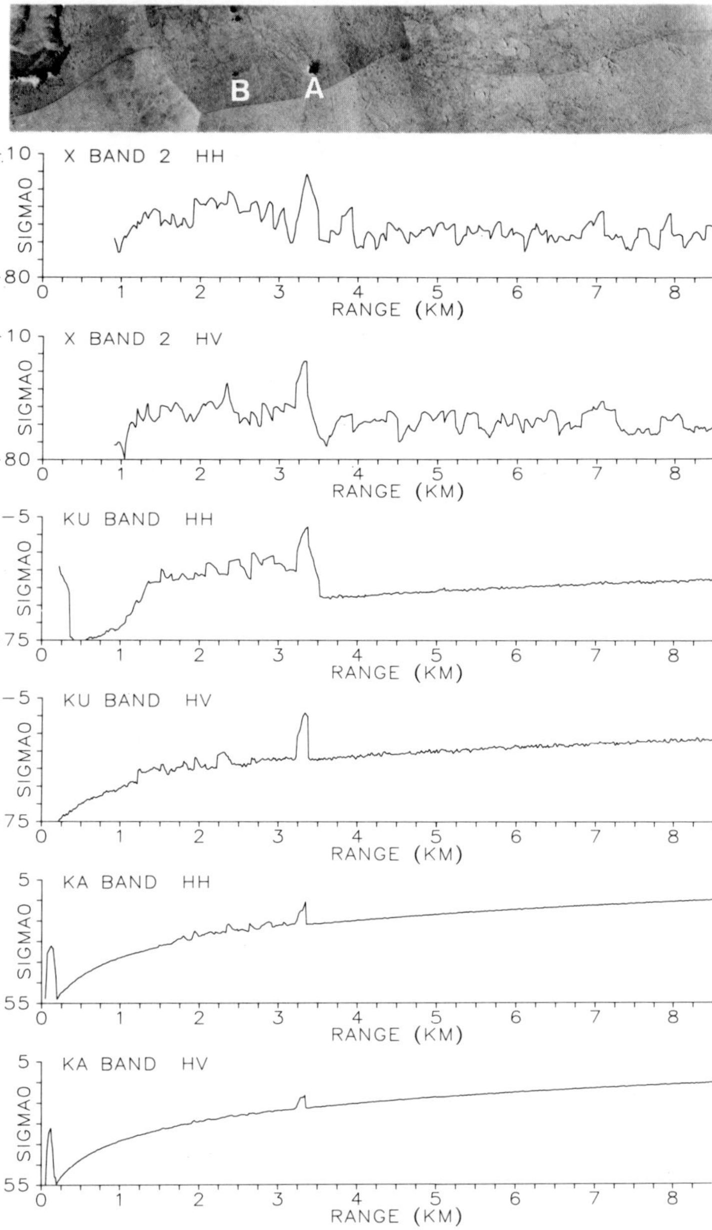

Fig 8.4 Photomosaic and like- and cross-polarized returns along Line 307 for X-, K_u-, and K_a-band. Targets of interest include the large iceberg at A and a smaller iceberg at B.

The second target, the small iceberg at B, was not detectable in the like-polarized returns with any of the frequencies, without prior knowledge of its location. The X-band like-polarized return had a sizable return at the correct range, but the return was distributed over a considerable range extent which did not identify a unique target. The X- and K_u-band cross-polarized returns, on the other hand, distinctly detected the small iceberg at B. The low K_a-band return (caused by low power transmission and lossy waveguide) did not provide enough cross-polarized energy to allow detection. Other examples of K_a-band returns, however, confirmed that the cross-polarized returns gave similar results to the X- and K_u-band radars.

For the subsequent analysis of polarization for the 1983 data, only the K_u-band results will be considered. There are two reasons for this. First, the K_a-band returns were too weak, and in many cases did not provide adequate signals for analysis. Second, the X-band radar employed a wide (3.5°) azimuthal beamwidth antenna which illuminated a large ice surface, particularly at far ranges, so that returns from targets well off the line of interest contaminated the returns under analysis. In addition, the X-band antenna had poor cross-polarized isolation which may have resulted in a leak of the like-polarized energy into the cross-polarized channel. The higher cross-channel isolation and the narrower beamwidth (1.5°) of the K_u-band antenna avoided these difficulties.

For the analysis of the 1983 data, only HH and HV returns are analyzed. This choice was justified based on the values of HH and VV, and HV and VH σ_0 shown in Table 8.2; this will be discussed later. The values of HV and VH are almost identical for the same targets of the ice types shown, illustrated diagrammatically in

Figure 8.5, a plot of HV and VH for Line 341. The HV and VH returns are almost indistinguishable. Although there is a larger variability in the HH and VV returns, there is no systematic difference. The H-transmit polarization was chosen because of its general usage.

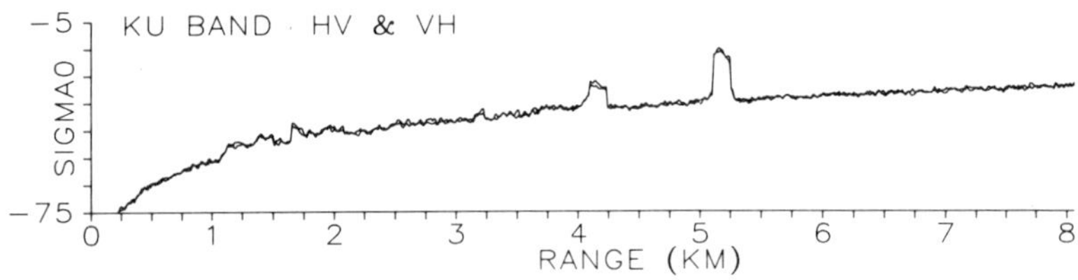

Fig. 8.5 Plot of K_u-band HV and VH returns for Line 341.

The top of Figure 8.6 again shows the normalized radar cross-section for like-polarized K_u-band returns along Line 307, the bottom shows the cross-polarized return, while the corresponding section of the photomosaic is shown in the centre. The target at A was a large iceberg, with a height of 45 m and a width of 75 m, yielding a physical cross-section of about 3,400 m². The very low signal levels behind the iceberg were a result of the radar shadow cast by the iceberg. Comparison of the like- and cross-polarized returns shows that the cross-polarized signals, with the exception of the return from the iceberg, were considerably lower than the like-polarized returns and made the iceberg more readily noticed in the cross-polarized plot. The target-to-clutter ratio for the large iceberg at A was 16 dB in the HH case and in the HV case increased to 26 dB, an improvement of 10 dB. Of even more significance, however, is the target at B. Although this target is not identifiable in the like-polarized plot on the top of

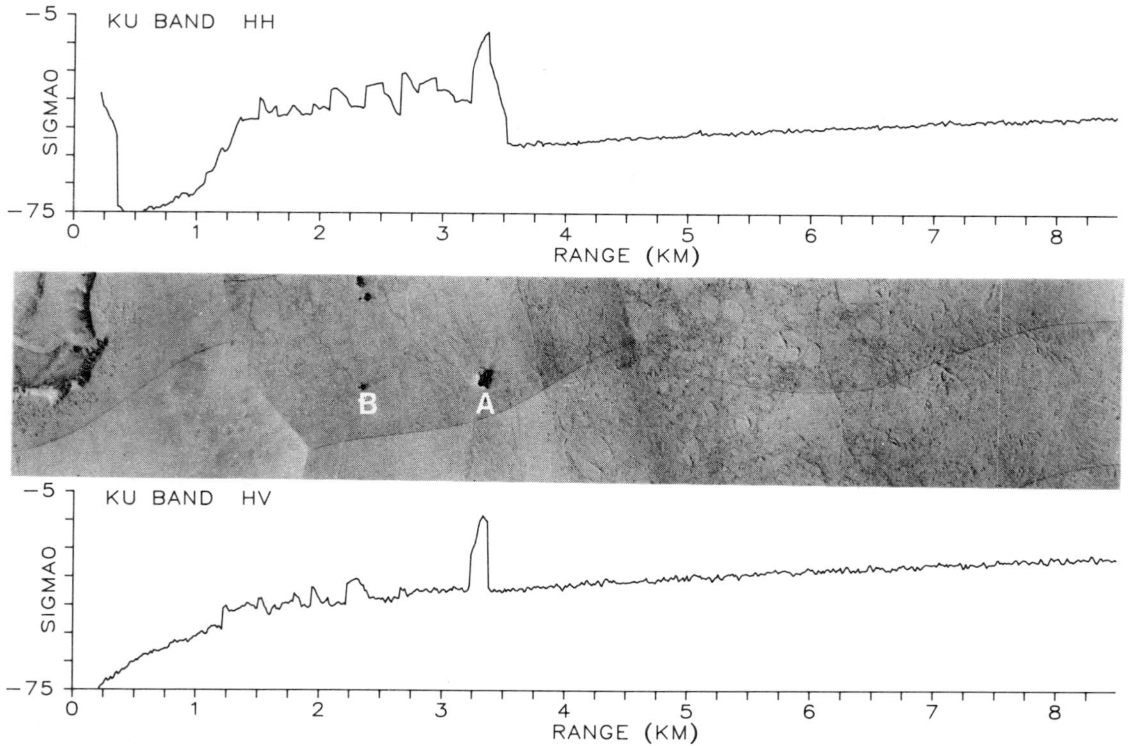

Fig. 8.6 Photomosaic and K_u-band like- and cross-polarized returns along Line 307. Targets of interest include the large iceberg at A and smaller iceberg at B.

this figure, it is well defined in the cross-polarized plot. Target B was another small iceberg (300 m^2) in front of the large iceberg.

For the small iceberg at B, the target was 8 dB below the clutter level in HH polarization, while in the HV polarization the iceberg was 4 to 9 dB above the surrounding clutter, an improvement in target-to-clutter ratio of between 12 and 17 dB. In this case, the cross-polarized return resulted in the detection of a hazardous target that was undetectable with the conventional HH polarization.

Another example of the use of cross-polarization in detecting targets of navigational interest is shown in Figure 8.7. The photomosaic and radar returns are for Line 341 from the 1983 data. The like-polarized K_u-band return is at the top, and the cross-polarized return is below the photomosaic. The targets of concern are the iceberg at B and the multiyear floe at A. The large iceberg (15,200 m^2) was well represented in both the like- and cross-polarized returns, although the iceberg was more clearly detectable in the cross-polarized response. There was, however, no target-to-clutter advantage for this target in the HV return.

The target at A, which was not distinguishable in the like-polarized return but was easily distinguishable in the cross-polarized reply, was a multiyear ice floe that would have imposed significant resistance to navigation. In the like-polarized return the multiyear floe gave the same σ_0 level as the surrounding clutter, while in the cross-polarized return the floe was 10 dB above the clutter, resulting in a 10-dB target-to-clutter improvement.

Another example of improvement in differentiation of targets is shown in Figure 8.8. The mosaic and radar returns correspond to Line 292 from the 1983 data. The target at B was a small iceberg that was not centrally

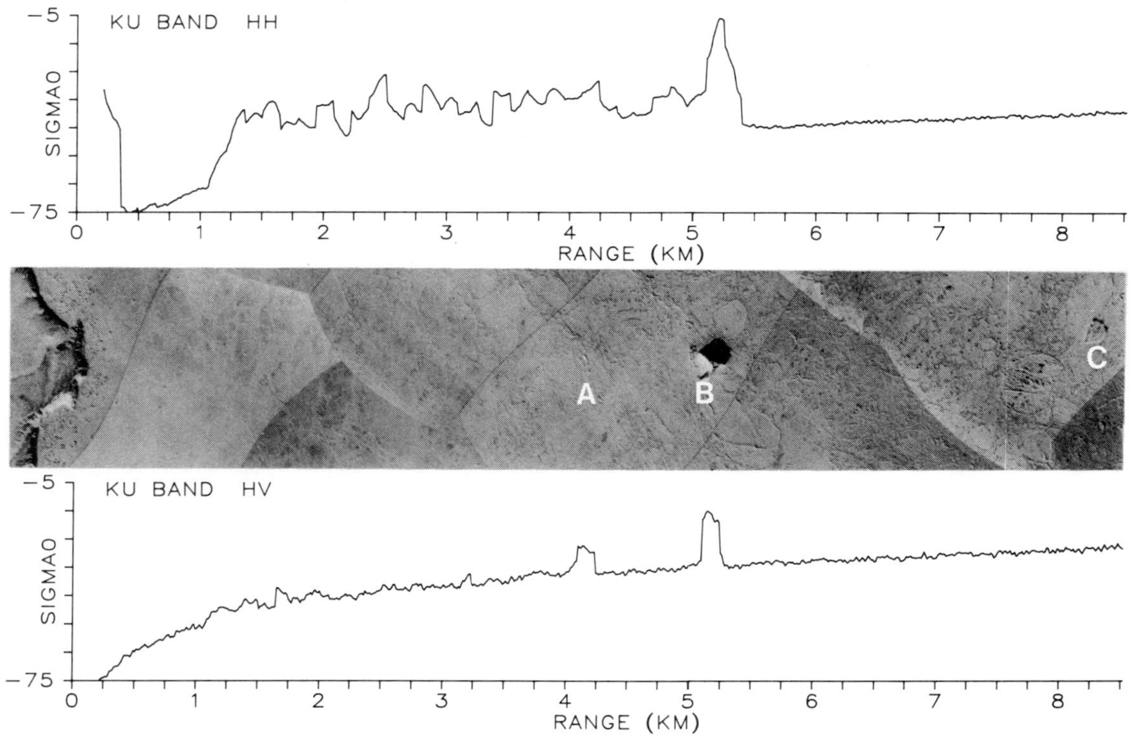

Fig. 8.7 Photomosaic and K_u-band like- and cross-polarized returns along Line 341. Targets of interest include multiyear ice floes at A and C and an iceberg at B.

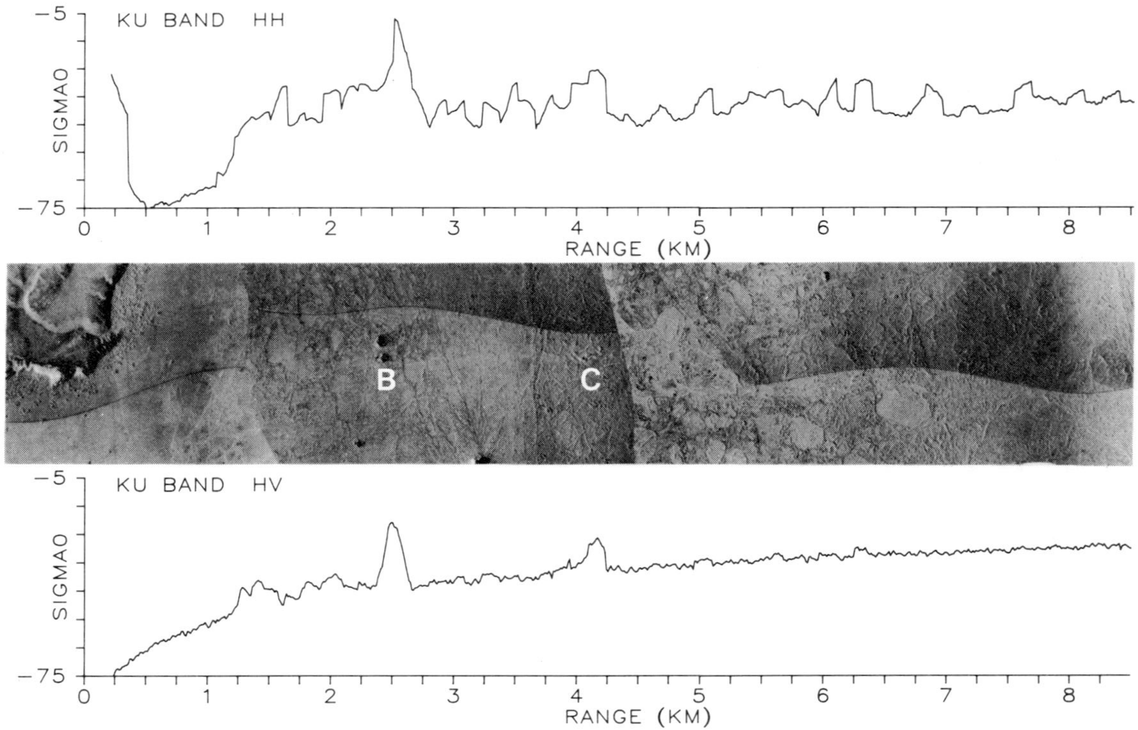

Fig. 8.8 Photomosaic and K_u-band like- and cross-polarized returns along Line 292. Targets of interest include an iceberg at B and multiyear ice floe at C.

located in the antenna beam. There was only a marginal (1 to 2 dB) target-to-clutter improvement for the iceberg in the HV return. The target at C was a small, but rough, multiyear floe. In the HH return, the multiyear floe σ_0 was 5 dB above the clutter (a first-year pressure ridge), while in the HV return the multiyear ice floe was 14 dB above the clutter, resulting in a 9-dB target-to-clutter improvement using cross-polarization.

A further example that demonstrates the benefit of employing the cross-polarized returns was shown in Figure 8.9 for Line 277. The target at A was a medium-sized, first-year pressure ridge, while the target at B was a multiyear ice floe. With the like-polarized return, the targets were almost indistinguishable. However, in the cross-polarized returns, the return from the first-year pressure ridge has been all but eliminated, while the return from the multiyear floe has only decreased 2 dB. The result was a 7-dB improvement in the ratio of the multiyear floe to the first-year pressure ridge, thereby making the multiyear floe clearly stand out.

The final example, the photomosaic and radar returns for Line 0, is shown in Figure 8.10. The target at A was a low-lying island (Adams Island), with a physical cross-section facing the radar of about 7,500 m^2. Although the island was easily detectable in the like-polarized return, there was a 10-dB target-to-clutter improvement in the cross-polarized display.

Table 8.1 summarizes the target-to-clutter improvement for the targets discussed and for several others.

Although X-band polarization results have not been shown because of difficulties with the X-band antenna, it is important to emphasize that the improvements resulting from using the cross-polarized returns also occur at X-band, although possibly to a somewhat lesser extent.

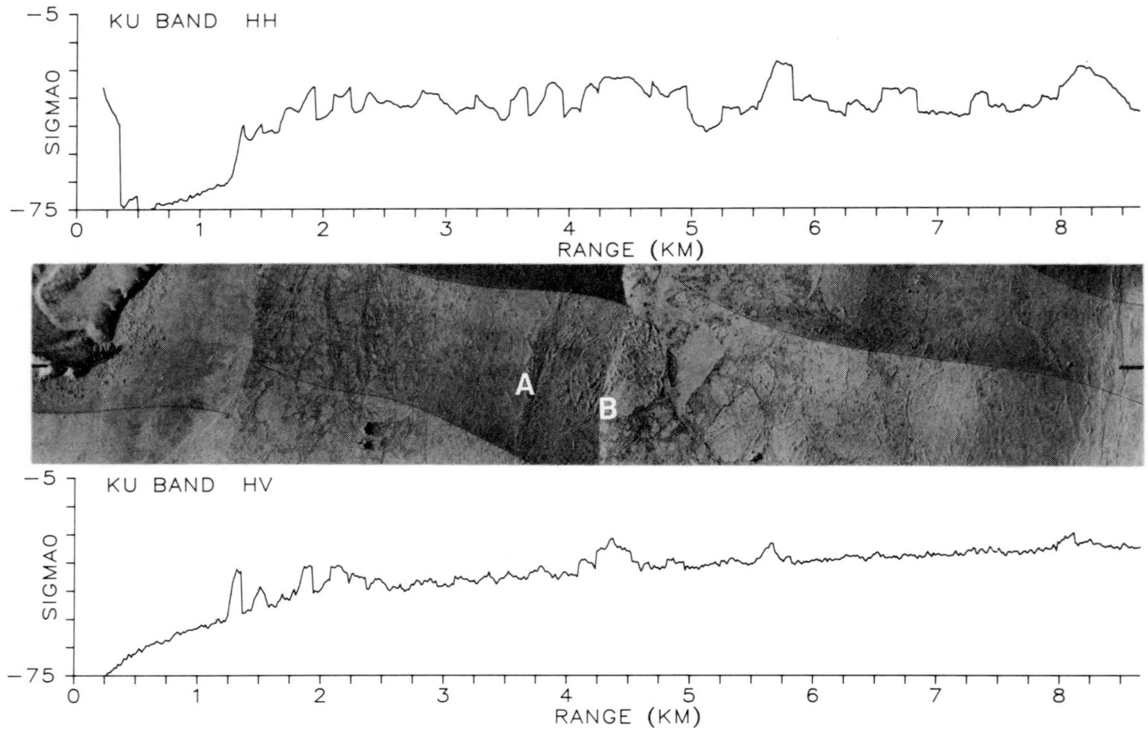

Fig. 8.9 Photomosaic and K_u-band like- and cross-polarized returns along Line 277. Targets of interest include a first-year pressure ridge at A and a multiyear ice floe at B.

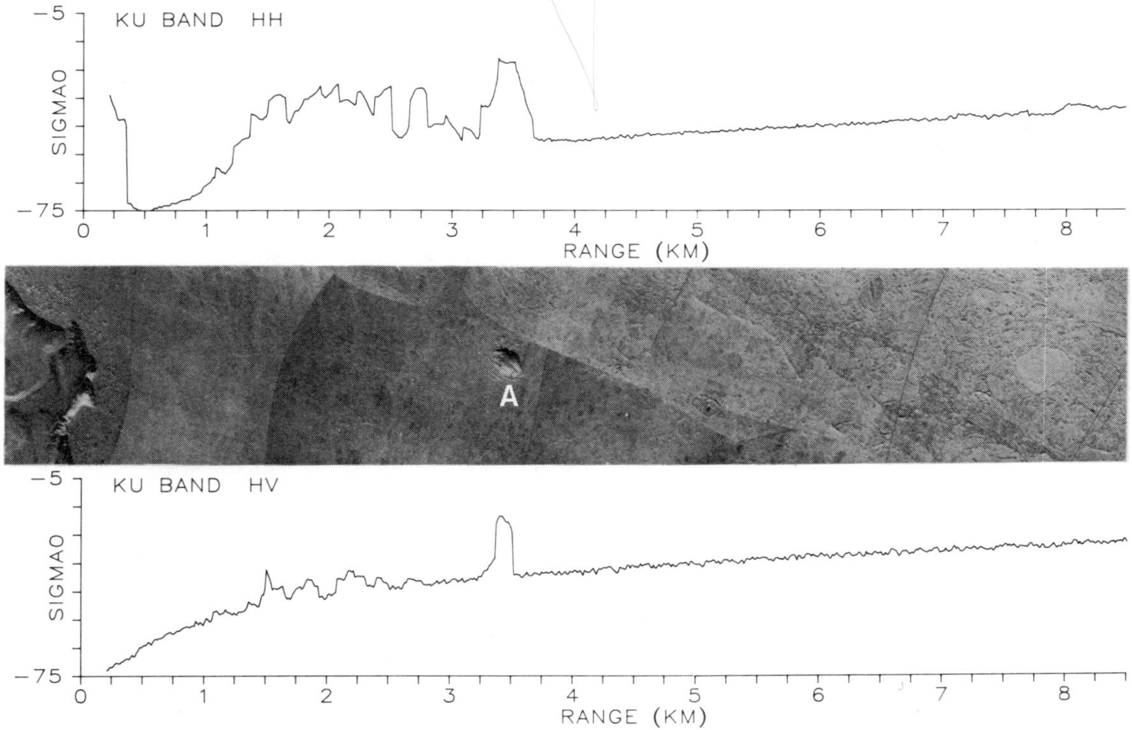

Fig. 8.10 Photomosaic and K_u-band like- and cross-polarized returns along Line O. Targets of interest include low-lying island (Adams Island) at A.

Table 8.1. Target-to-clutter improvement for selected targets using cross-polarized returns.

Target	Line	T/C improvement (dB)
Iceberg A	307	10
Iceberg B	307	12 to 17
Iceberg D	341	-3*
Multiyear floe C	341	10
Iceberg B	292	1
Multiyear floe C	292	9
Multiyear floe B	277	7
Adams Island B	0	10

* T/C ratio reduced.

Figure 8.11 shows the X-band results for Line 307. The target at A was the large iceberg, while the target at B was the small iceberg. The cross-polarized return provided a 6-dB target-to-clutter improvement for the large iceberg, and a 7-dB target-to-clutter improvement for the smaller iceberg. These improvements would be even more substantial if a better antenna were employed.

The improvements in detectability of desired ice targets discussed in the previous examples can be clearly observed in image format, as shown in Figure 8.12. The images were generated on a high-resolution raster scan display system and photographed in the laboratory. The image at the top of the figure shows the K_u-band like-polarized returns, while the image on the bottom shows the cross-polarized returns.

The large iceberg (1) on Line 307 was visible on both images, while the smaller iceberg (2) closer to the radar was not visible in the like-polarized image. The cross-polarized returns showed the large iceberg clearly visible at 1, while the smaller iceberg at 2 was also

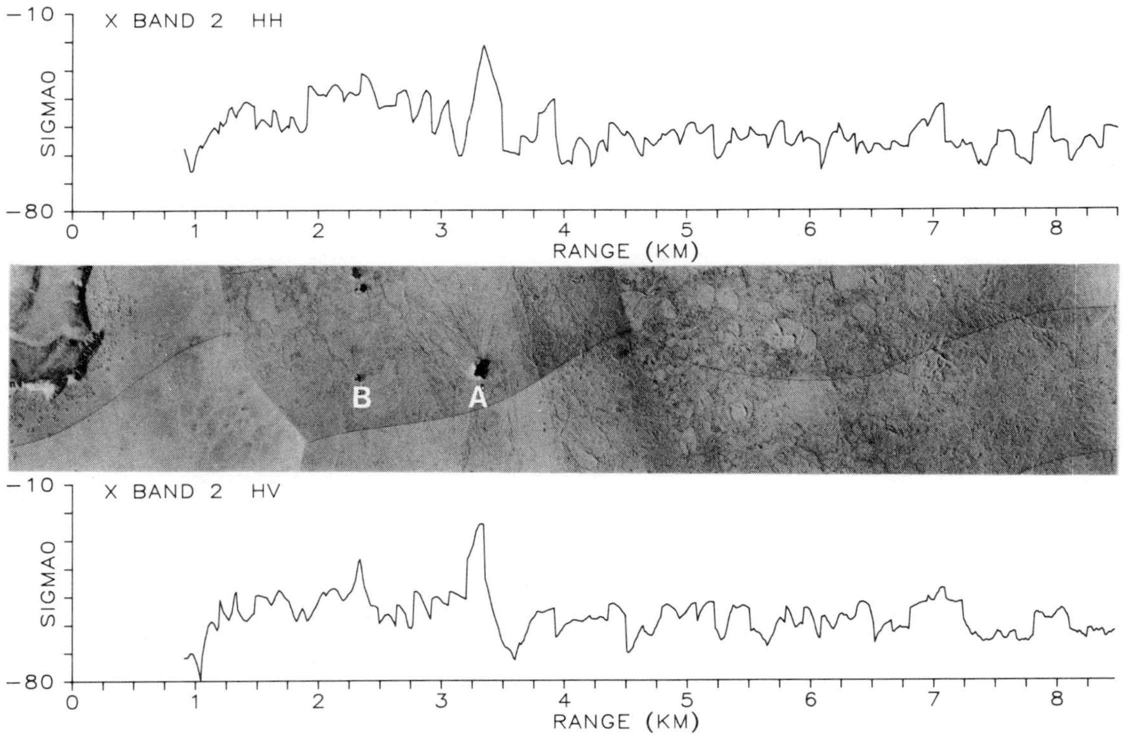

Fig. 8.11 Photomosaic and X-band like- and cross-polarized returns along Line 307.

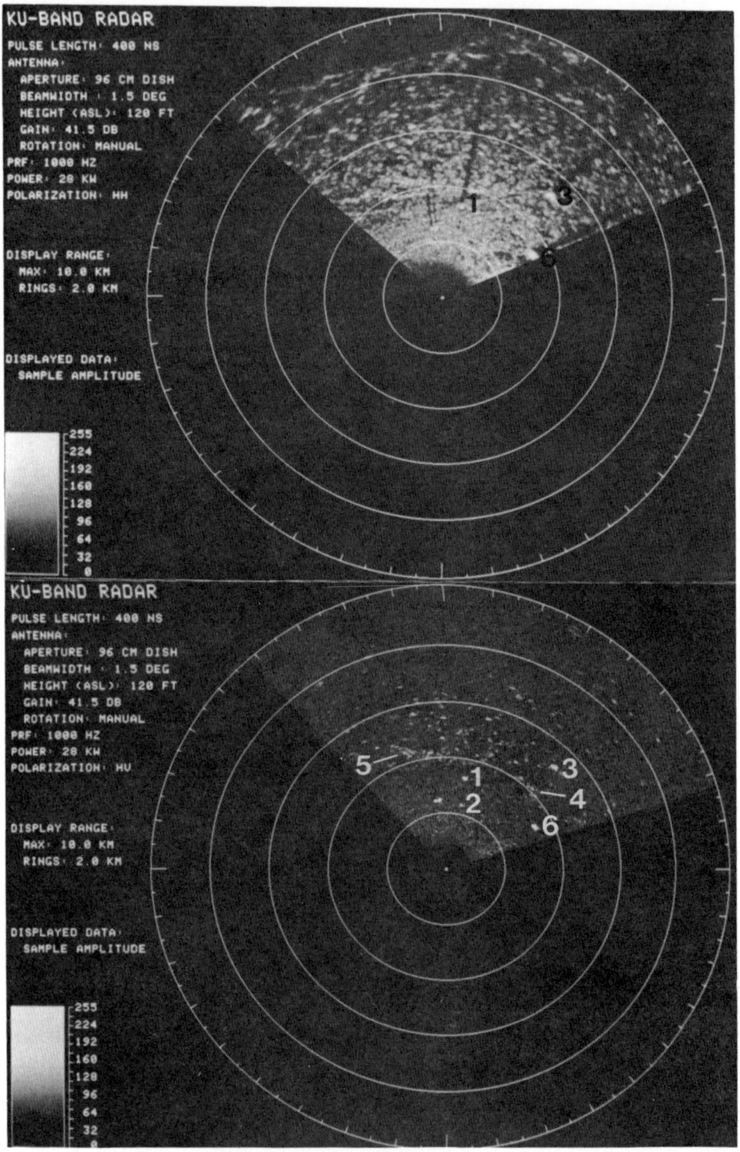

Fig. 8.12 K_u-band like-polarized (top) and cross-polarized (bottom) images. Targets include: 1) large iceberg Line 307; 2) small iceberg Line 307; 3) iceberg Line 341; 4) multiyear floe Line 341; 5) multiyear floe Line 277; 6) island.

detectable. The iceberg (3) on Line 341 was identifiable on both the like- and cross-polarized image, although it was more clearly discernible on the cross-polarized image. However, the multiyear floe (4) on Line 341 was not visible on the like-polarized image but was identifiable on the cross-polarized image. The multiyear floe (5) on Line 277 was not identifiable or separable from the first-year pressure ridge on the like-polarized image but was clearly identifiable on the cross-polarized image. Although the island (6) on Line 0 was visible on the like-polarized return it was more clearly detectable on the cross-polarized image. In general, there were significantly fewer targets on the cross-polarized image, and of the ones that were displayed, a good number were known to be either iceberg or multiyear ice. No known targets of interest (icebergs or multiyear ice) were absent from the cross-polarized image.

Although the K_a-band results have not, in general, been displayed because of the low signal levels associated with that frequency, it is worthwhile to examine the cross-polarized image for K_a-band operation and to compare this to the cross-polarized image for K_u-band operation shown in Figure 8.12. The K_a-band image (Figure 8.13) clearly identified the icebergs at 1, 2, and 7 and the island at 6. The iceberg at 3 was poorly displayed because of the drop-off in returned power with range. Also, as the antenna was manually scanned from northwest to southeast the main vertical lobe of the antenna was focused closer and closer in range, resulting in a further decrease in illumination for distant targets in the northeast and east sections of the image. However, several multiyear ice floes, particularly at 8, 9, and 10, were displayed more clearly on the K_a-band cross-polarized image of Figure 8.13 than in the K_u-band cross-polarized image of Figure 8.12.

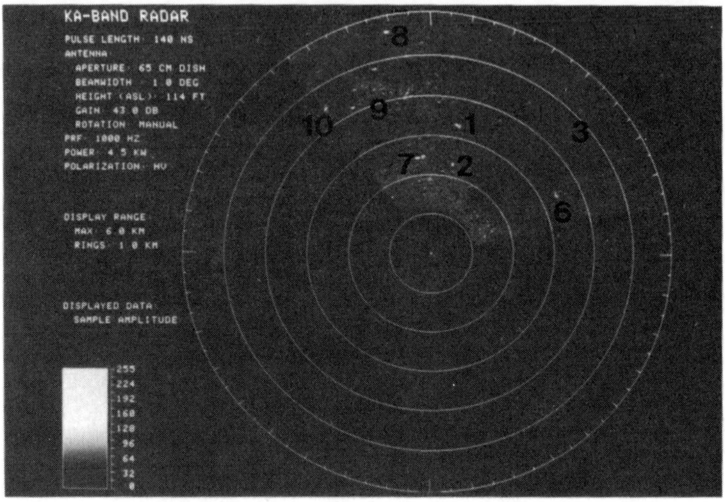

Fig. 8.13 K_a-band cross-polarized image. Targets include: 1) large iceberg Line 307; 2) small iceberg Line 307; 3) iceberg Line 341; 6) island; 7) icebergs; 8), 9) and 10) multiyear ice floes.

The strong depolarization of the radar signal by icebergs and multiyear ice results from the same phenomenon. Volume scattering, caused by air inclusions and irregularities in the ice, depolarizes the incident radar wave to a much larger extent for iceberg and older ice than for first-year ice. In icebergs, the air inclusions result from air entrapped during compression of the snow to form glacial ice, while the irregularities include pressure cracks and ice faces. In the case of multiyear ice, the air inclusions are largely a result of recrystallization of the ice during the summer melt periods. This volume scattering phenomenon relies on some vertical relief in the ice, thus permitting penetration by the radar wave. Icebergs typically present significant vertical relief, whereas multiyear floes have variable surface relief. However, the greater the relief, the greater the depolarizing effect. Alternately, the higher the antenna, the larger the grazing angle and, similarly, the greater the depolarizing effect.

Table 8.2 summarizes the σ_0 values, the ratios between HH and VV, and the ratio between the like- and cross-polarized components for the targets. The table shows that icebergs have the highest like-polarized and cross-polarized cross-sections, which result from their greater vertical relief. The target with the next highest σ_0 value both in like- and cross-polarization is the island. Multiyear ice has the third strongest like- and cross-polarized σ_0 values, while rough first-year ice and then moderately rough first-year ice have the lowest σ_0 values. On average, the values of horizontal like-polarized σ_0 for the various targets are -7.8 dB for iceberg ice, -21.0 dB for the island, -27.5 dB for multi-year ice, -28.4 dB for rough first-year ice, and -28.7 dB for moderately rough first-year ice. The average σ_0 values for the last three categories span only 1.2 dB, which supports the difficulty, as shown in the figures earlier, of differentiating multiyear ice from first-year ice. In contrast, the cross-polarized σ_0 values give average values of -16.2 dB for the icebergs, -18.0 dB for the island, -27.0 dB for multiyear ice, -37.1 dB for rough first-year ice, and -42.0 dB for moderately rough first-year ice. The multiyear ice σ_0 is on average 10 dB higher than the rough first-year ice which is, in turn, 5 dB stronger than the moderately rough first-year ice. As shown in the figures previously discussed, this substantially aids in the differentiation of these ice types.

Another interesting observation to be made from the data in Table 8.2 is the degree of depolarization caused by each target type. Columns HH/HV and VV/VH show the ratio of the like- to cross-polarized σ_0 for horizontal

Table 8.2 Normalized radar cross-sections at K_u-band for various ice features (1983).

Feature	Radar cross-section (dB) HH	VV	HV	VH	HH/VV	HH/HV	VV/VH	Average like/cross
Iceberg								
Line 292	-6.0	-7.3	-20.0	-20.0	1.3	14.0	12.7	
Line 307	-11.0	-15.0	-13.5	-13.5	4.0	2.5	-1.5	
Line 341	-6.3	-12.5	-15.0	-15.0	6.2	8.7	2.5	
Mean	-7.8	-11.6	-16.2	-16.2	3.8	8.4	4.6	6.5
Multiyear ice								
Line 252	-27.5	-27.5	-23.7	-25.0	0	-3.8	-2.5	
Line 247								
(1)	-30.5	-28.0	-23.5	-23.5	-2.5	-7.0	-4.5	
(2)	-22.0	-28.0	-24.5	-24.5	6.0	2.5	-3.5	
Line 277								
(1)	-27.5	-27.5	-26.0	-26.0	0	-1.5	-1.5	
(2)	-31.5	-26.5	-34.0	-34.0	-5.0	2.5	7.5	
(3)	-22.0	-24.5	-28.0	-28.0	2.5	6.0	3.5	
(4)	-25.0	-18.5	-24.5	-24.5	-7.0	-0.5	6.0	
Line 292	-26.0	-22.0	-26.3	-26.3	-4.0	0.3	4.3	
Line 319								
(1)	-27.5	-27.5	-30.0	-30.0	0	2.5	2.5	
(2)	-34.5	-29.5	-30.0	-30.0	-5.0	-4.5	0.5	
Line 341	-28.5	-25.0	-27.0	-27.0	-3.5	-1.5	2.0	
Mean	-27.5	-25.9	-27.0	-27.2	-1.7	-0.5	1.3	0.4
First-year ice								
Line 252								
Ridge	-28.5	-31.0	-37.5	-40.0	3.5	9.0	6.5	
Rough	-25.0	-25.5	-36.0	-38.0	0.5	11.0	12.5	
Line 247								
Ridge	-30.0	-32.5	-34.0	-34.0	2.5	4.0	1.5	
Line 319								
Ridge	-30.0	-28.0	-37.0	-37.0	-2.0	7.0	9.0	
Line 277								
Ridge	-30.0	-30.0	-36.0	-36.0	0	6.0	6.0	
Line 341								
Ridge	-27.0	-30.0	-42.0	-42.0	3.0	15.0	12.0	
Mean	-28.4	-29.5	-37.1	-37.8	1.3	8.7	7.9	8.3
Line 341								
Mod. rough	-30.0	-30.0	-44.0	-44.0	0	14.0	14.0	
Line 247								
Mod. rough	-32.0	-32.0	-42.0	-42.0	0	10.0	10.0	
Line 307								
Mod. rough	-24.0	-24.0	-40.0	-40.0	0	16.0	16.0	
Mean	-28.7	-28.7	-42.0	-42.0	0	13.3	13.3	13.3
Island								
Line 0								
Adams Island	-21.0	-14.0	-18.0	-20.0	-7.0	-3.0	6.0	1.5

and vertical transmit. The average of the cross-polarized returns shows that the multiyear ice was the strongest depolarizer, with the like polarization on average having a σ_0 of only 0.4 dB above the cross polarization. The island was the next strongest depolarizer, with the like-polarized σ_0 being 1.5 dB above the cross-polarized σ_0. The icebergs were third, with a like-to-cross ratio of 6.5 dB, while the rough and moderately rough first-year had ratios of 8.3 dB and 13.3 dB, respectively.

The final point to make with respect to Table 8.2 concerns the ratio of the horizontal transmit/horizontal receive (HH) to vertical transmit/vertical receive (VV), i.e., HH/VV. With the exception of the returns from the island (which favoured VV by 7 dB), there was little to choose between HH or VV, although the iceberg returns favoured the HH combination by 3.8 dB. This result suggests that the conventional choice of horizontal transmit is also the most appropriate for surface-based sensing of ice.

The results tabulated in Table 8.2 can be demonstrated more graphically as shown in Figure 8.14. The cross-polarized, normalized radar cross-section, σ_{0HV}, is plotted versus ice hazard. It is clear that the lowest values are associated with moderately rough first-year ice, while the highest values are associated with icebergs. In addition, there is a linear increase in σ_{HV} from moderately rough first-year, to rough first-year, to moderately rough multiyear ice, up to the highest values associated with the icebergs. This effect can be further illustrated as shown in Figure 8.15, where the cross-polarized, normalized radar cross-section, σ_0, is plotted against surface relief. For equivalent surface relief, the multiyear ice σ_{0HV} values are 12 to 14 dB higher than the first-year ice values.

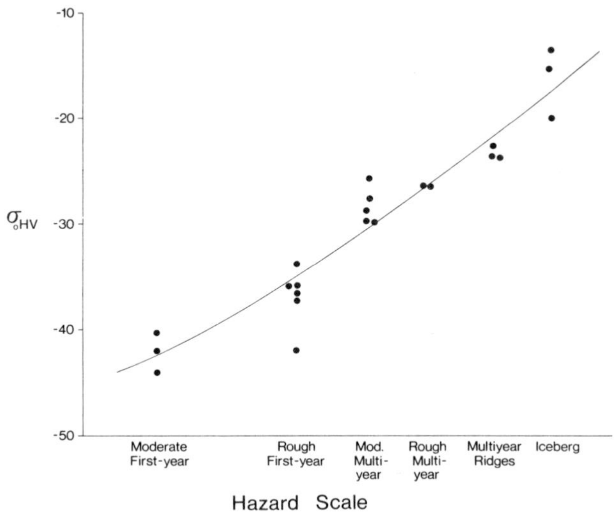

Fig. 8.14 Plot of K_u-band cross-polarized returns, σ_oHV, versus hazard scale.

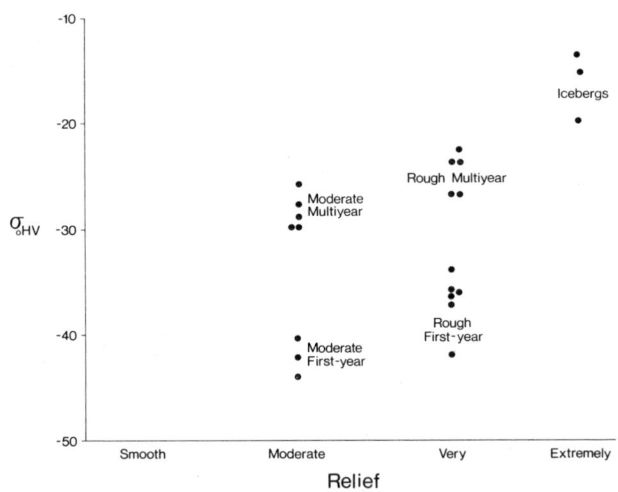

Fig. 8.15 Plot of K_u-band cross-polarized returns, σ_oHV, versus surface relief.

Finally, a plot of the like- versus cross-polarized σ_o values can be constructed, as shown in Figure 8.16. It shows that a classification system can be established which would distinguish between icebergs and multiyear or first-year ice, and between multiyear ice and smooth or moderately rough first-year ice. The separation between rough first-year ice and moderately rough multiyear ice is not as substantial as that between the other ice types. In classifying ice as multiyear or first-year based on the σ_{oHV} value, it would appear better to err on the side of caution and declare any target with σ_{oHV} greater than about -35 dB as multiyear.

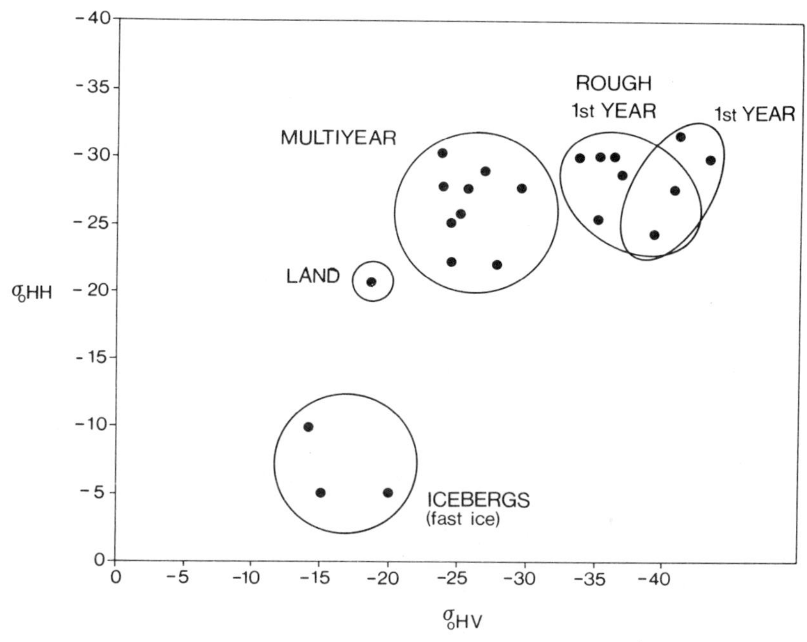

Fig. 8.16 Plot of K_u-band like-polarized (HH) σ_o, versus cross-polarized (HV) σ_o.

<u>Open Water Results</u>

Whereas the ice targets examined in the previous sections were locked in a stationary ice cover during the field programs, many of the ice targets examined in the 1984 experiment in open water conditions experienced significant motion and aspect change during the data collection. It was expected that the targets whose aspect changed with respect to the radar would experience large changes in returned power, and that smaller and lower targets would fluctuate because of multipath interference. However, all the targets, including several grounded icebergs which experienced little or no aspect change with respect to the radar, experienced large variations in return power. In addition, the variability occurred with a periodicity such that the changes could not be ascribed to actual physical motion of the target.

For example, target B22, Figure 8.17 (a small, grounded iceberg with about 400 m^2 of surface exposed to the radar) had a variability of up to 14 dB in the HH return

Fig. 8.17 Photo of ice target B22, with surface-truth boat to the right.

and 10 dB in the HV return (Figure 8.18) within a 30-s sampling period. The 128 samples shown were collected at equally spaced intervals over the 30-s sample period. Over a four-day period the HH return varied 14 dB and the HV varied 24 dB. The short-term (30-s) variation was 9 dB for the VV return and 8 dB for the VH return. The four-day variability was 10 dB for the VV return and 9 dB for the VH return. This target exhibited no noticeable change in aspect with respect to the radar during the experiment and was high enough (15 m) to minimize the expected effects of multipath interference. The target was at a range of only 2,500 m, and should not have been affected by any propagation effects from changing atmospheric conditions. In fact, the weather and the sea state were essentially stable during the experimental program. All the measurements during the 1984 program were made with a dual-polarized X-band radar system.

The difference between the like- and cross-polarized returns was just as variable as the returns themselves,

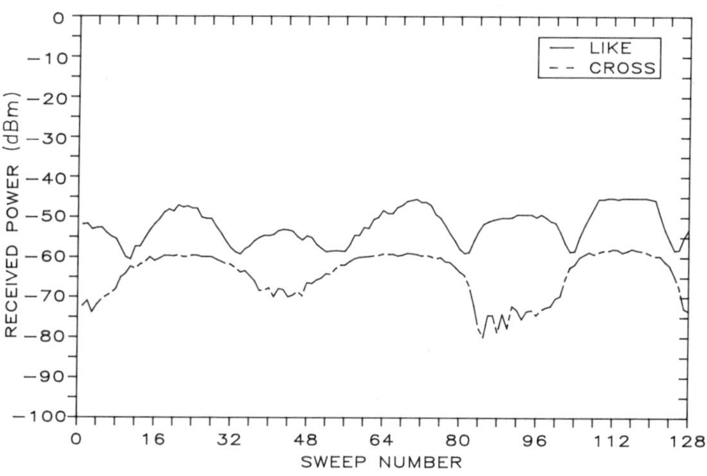

Fig. 8.18 Variability in like- and cross-polarized radar returns from target B22, over a 30-s period. Transmitted polarization was horizontal.

with the like-to-cross ratio changing from 7 to 25 dB within one data set (30 s) to a range of -1 (stronger cross return) to +8 dB within another set on another day.

Another important observation from the experiment was that all targets significantly depolarized the radar signal, although with a high variability. This depolarizing effect was also shown in the previous results for the stationary ice environment. As the dual-polarized dish antenna used for this experiment had a cross-polarized isolation of better than 25 dB, like-to-cross ratios of less than 25 dB can be attributed to depolarization caused by the target.

For the larger, stationary targets, the like-to-cross ratio varied from -5.5 to +37.0 dB. An analysis of returns from these targets shows that the average like-to-cross ratio was 8.4 dB. There seems to be little dependence on transmitter polarization, as the average HH/HV ratio was 8.2 dB, while the average VV/VH ratio was 9.7 dB.

To acquire statistics on returns from the larger stationary targets, all data from four large targets (B22, B23, B26, B27) were accumulated over five days. The radar equation was used to convert the return power to target radar cross-section. This cross-section was then normalized by the target's physical area, as seen by the radar, to yield the normalized radar backscatter coefficient, σ_0, expressed in dB with respect to 1 m^2. The cross-section was normalized with only the target area (not the entire radar cell) since the surrounding ocean did not contribute any significant return. Figure 8.19 shows the histogram of σ_0 values for like- and cross-polarizations for target B22. The range of like-polarized σ_0 values covered 27 dB with an average value of -20.0 dB. The like-polarized histogram for

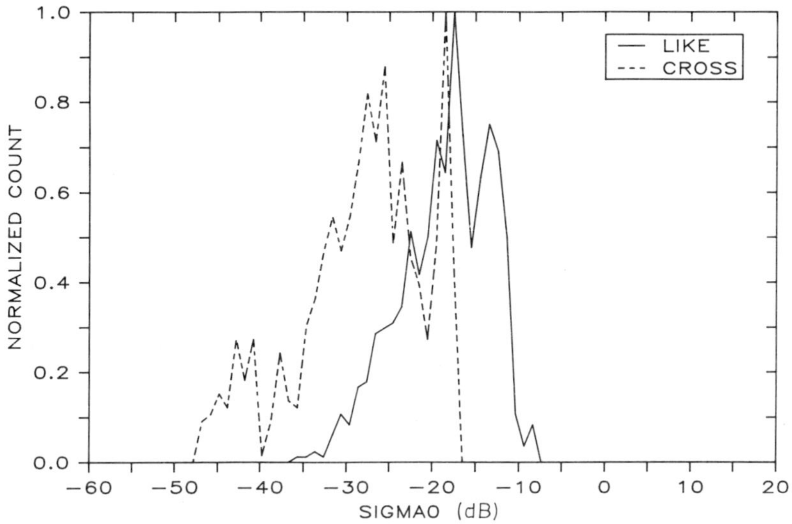

Fig. 8.19 Histograms of like- and cross-polarized radar cross-sections for target B22.

target B23 (Figure 8.20) had a range of 26 dB, with a higher average σ_0 of -10.3 dB.

The other two larger targets (B26 and B27) had similar spreads of σ_0 values and yielded like-polarized averages of -11.5 dB (B27) and -7.0 dB (B26). The lower σ_0 value associated with B22 probably resulted from the flat smooth face of the target sloping away from the radar.

To determine the ice target sensitivity to transmitter polarization, separate HH and VV histograms were accumulated. Targets B23 and B27 had average VV σ_0 values of 5.1 and 3.4 dB, respectively, above the HH σ_0 values; conversely, target B22 had an average HH σ_0 value that was 2.6 dB above the VV σ_0 value, while B26 exhibited no difference. Regardless of polarization, the like-to-cross ratio of σ_0 ranged between 6.4 and 10.2 dB for the four targets, with an average of 8.5 dB.

A representative sample of the smaller targets showed an HH/HV ratio of about 8.5 dB, while the VV/VH ratio was about 8.7 dB.

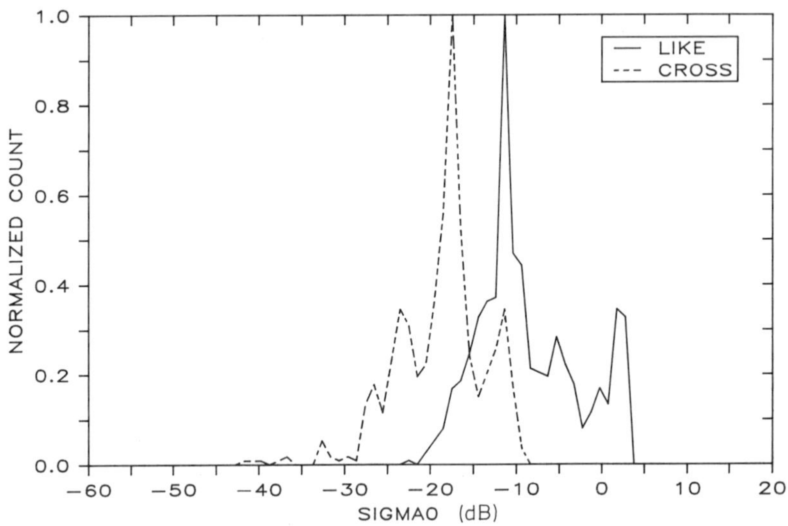

Fig. 8.20 Photo of target B23 (with surface-truth boat to the left) and histograms of like- and cross-polarized radar cross-section.

One of the smaller targets, B33, (with physical cross-section of about 11 m²) for which a large number of data samples was collected, was analyzed to provide statistics on returns from small non-stationary targets. Figure 8.21 shows the target and the histograms of like- and cross-polarized σ_0 values for all data collected for target B33. The average like-polarized σ_0 value for all returns was -15.1 dB, while the range of σ_0 values was from -37 to +2 dB. There was little difference between the average like-polarized σ_0 value for HH (-15.2 dB) and VV (-14.9 dB). In addition, the cross-polarized σ_0 values for HV and VH were within 0.8 dB, while the average like-to-cross ratio was about 7.5 dB. Table 8.3 summarizes σ_0 values for all of the targets.

In reviewing the 1984 data, it became apparent that although all targets occasionally returned a cross-polarized signal stronger than the like-polarized return, this phenomenon happened more frequently with the smaller targets. Target B28, a growler of about 26 m², gave cross-polarized returns greater than 13 dB above the like-polarized returns on several sample sets. Figure 8.22 plots the received power for 128 samples collected at uniform intervals over a 30-s period. The like-polarized return is plotted with the solid line, while the cross-polarized return is plotted with the dotted line. This situation of substantial signal advantage for the cross-polarized return lasted for periods of up to several minutes. Similarly, target B31 (about 1.5 m²) gave cross-polarized returns up to 10 dB stronger than the like-polarized returns. This higher cross-polarized return often permitted target detection, even though the like-polarized return was lost in clutter or noise.

Of particular interest is the observation that although the like- and cross-polarized returns tended to vary

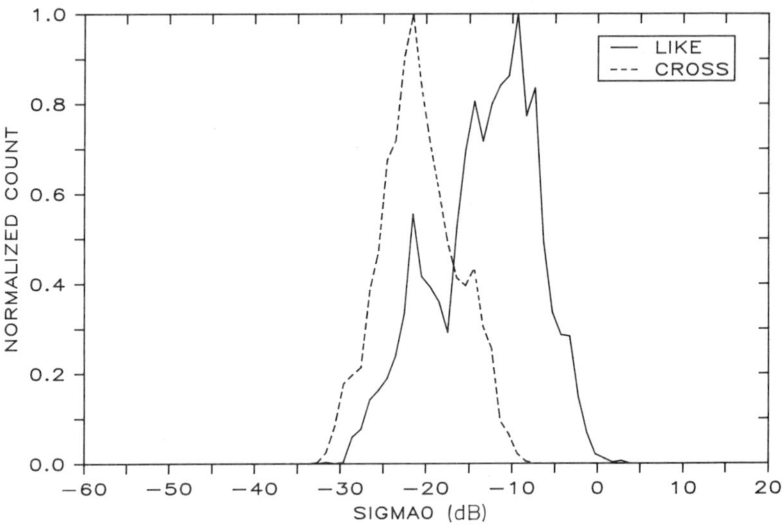

Fig. 8.21 Photo of target B33 (with 0.5 m buoy to the right) and histograms of like- and cross-polarized radar cross-section (σ_o).

Table 8.3. Normalized radar cross-sections at X-band for selected ice targets (1984).

Target	Cross-section (m^2)	σ_0 like	σ_0 cross	σ_0 HH	σ_0 HV	σ_0 VV	σ_0 VH	Like/ cross	HH/ HV	VV/ VH
B16	1.8	-15.8	-24.1	-15.8	-24.1			6.3	6.3	
B17	5.0	-6.8	-15.1	-6.8	-15.1			7.0	7.0	
B18	12.0	-10.2	-19.7	-10.2	-19.7			7.5	7.5	
B19	21.0	-10.2	-20.6	-10.2	-20.6			8.0	8.0	
B20	19.0	-13.7	-22.1	-13.7	-22.1			6.7	6.7	
B21	260.0	-19.1	-23.6	-18.7	-23.3	-22.8	-29.4	3.8	3.7	5.2
B22	500.0	-20.0	-26.7	-19.5	-25.6	-22.1	-32.7	6.4	5.7	9.7
B23	320.0	-10.3	-20.8	-11.0	-21.5	-5.9	-17.3	10.2	10.0	11.3
B24	82.0	-8.3	-22.7	-8.5	-22.8	-5.1	-17.0	10.5	10.4	14.9
B25	300.0	-16.0	-24.2	-16.8	-24.9	-9.4	-18.3	8.0	7.9	8.6
B26	1200.0	-7.0	-15.6	-6.9	-15.0	-7.3	-17.3	9.0	9.0	9.2
B27	3200.0	-11.5	-20.3	-12.5	-22.3	-9.1	-17.4	8.4	9.5	6.0
B28	55.0	-14.2	-19.4	-14.2	-19.4	-14.2	-19.3	4.4	4.4	4.4
B30	6.0	-13.7	-18.3	-13.7	-18.3			3.5	3.5	
B31	1.5	-20.8	-29.8	-22.6	-29.1	-19.8	-30.1	8.3	7.0	9.8
B33	11.1	-15.1	-22.9	-15.2	-23.2	-14.9	-22.4	7.5	8.2	6.3
B34	8.0	-9.7	-21.2	-13.5	-20.4	-8.4	-21.4	9.4	7.7	9.9
B35	9.0	-15.1	-24.1	-15.1	-24.1			8.3	8.3	
B36	260.0	-24.3	-28.9	-28.5	-31.9	-15.4	-22.2	6.5	5.6	7.4
B37	3900.0	-8.5	-21.4	-8.1	-18.1	-9.7	-31.2	11.8	9.5	20.5
B38	22.0	-9.8	-23.2	-9.8	-23.2			12.4	12.4	
B39	17.0	-8.2	-21.6	-8.2	-21.6			13.7	13.7	
B40	6.0	-12.0	-23.4	-12.0	-23.4			12.1	12.1	
B41	15.0	-15.1	-19.6	-15.1	-19.6			5.2	5.2	
B43	69.0	-14.9	-25.6	-14.9	-25.6			9.0	9.0	
B44	40.0	-23.4	-29.0	-23.4	-29.0			5.7	5.7	
B45	13.0	-21.8	-31.6	-21.8	-31.6			10.5	10.5	
B48	4.0	-8.2	-21.3	-8.0	-21.3			13.8	13.8	
B49	10.0	-12.0	-25.7	-12.0	-25.7			13.1	13.1	
Average		-13.64	-22.84	-14.03	-22.84	-12.62	-22.77	8.52	8.32	9.48

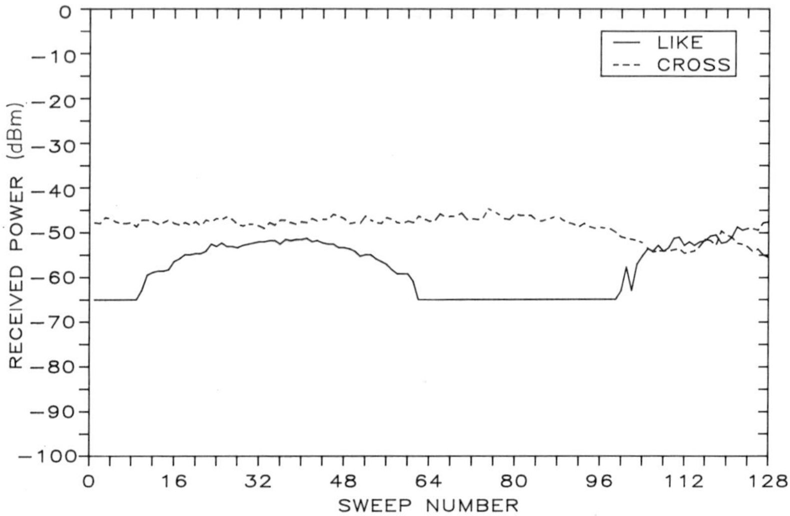

Fig. 8.22 Variability of like- and cross-polarized radar returns from target B28, collected over a 30-s period. Transmitted polarization was horizontal.

considerably with time, these variations were not in phase, as shown in Figure 8.23. When the like-polarized return was reaching a minimum, the cross-polarized return was often high, and vice versa. Examination of many other sets of data confirms the frequent independence of the like- and cross-polarized returns. This implies the possibility of improved target detection through the combined use of like- and cross-polarized signals.

This improved detection can be visualized by reference to Figure 8.24, which plots the like- and cross-polarized returned power from a small ice target for 128 consecutive sweeps. With an arbitrary threshold level of -60 dB (as shown in the figure), a target would be declared 59% of the time in the like-polarized channel. Using only the cross-polarized channel, a target would be declared 45% of the time, while if both channels were employed, a

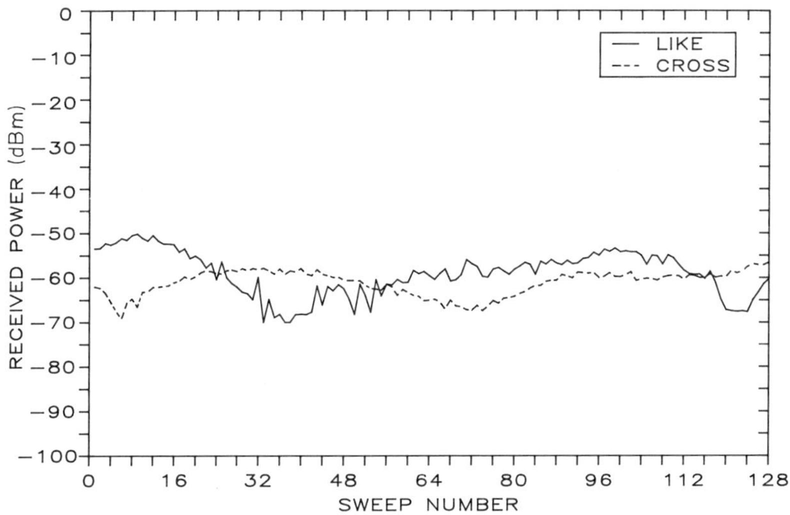

Fig. 8.23 Variability of like- and cross-polarized radar returns from a small ice target, collected over a 30-s period. Transmitted polarization was horizontal.

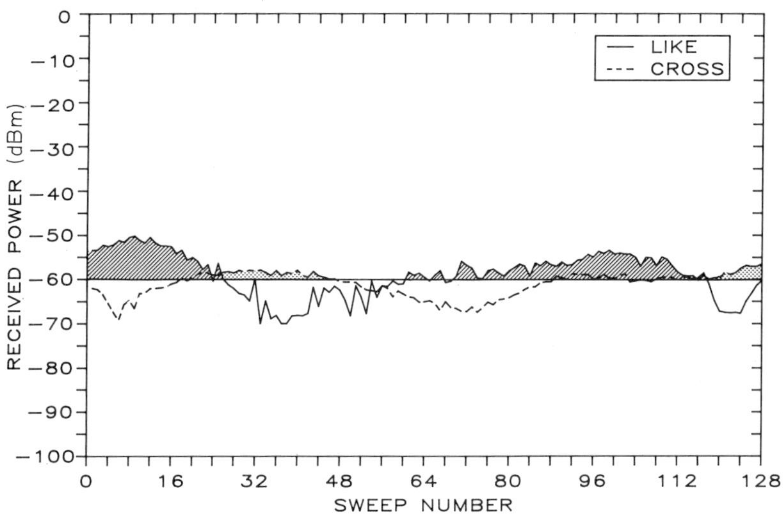

Fig. 8.24 Demonstration of improved probability of detection by using both like- and cross-polarized channels. With threshold of -60 dBm, detection occurs 59% of the time with only like-polarized channel, 45% of the time with cross-polarized channel, but 81% of the time by using both channels.

target would be declared 81% of the time. The use of both channels has improved the detection by 22%.

For the smaller ice targets, it is interesting to note that there was a strong correlation between the frequency of variation of the returns from the target and from the sea clutter surrounding it. Figure 8.25 shows a situation where the period of the target oscillations in (b) is the same as the sea clutter oscillations in (a). This situation probably occurred because the target was not only being vertically displaced by the waves at the wave frequency (and thus the sea clutter frequency), but was also being partially obscured by the waves, causing target return scintillation.

When the like- and cross-polarized σ_0 values for icebergs from the open-water experiment are plotted with the values obtained from the fixed ice situation, the result is as shown in Figure 8.26. It must be remembered, however, that the open-water data were collected at X-band,

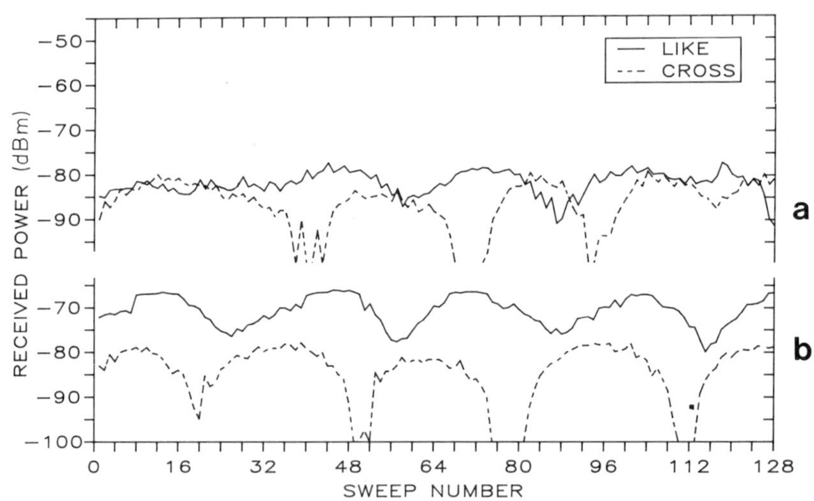

Fig. 8.25 Correlation between variations in sea clutter (top traces) and small ice target returns (bottom traces). Like polarization is HH.

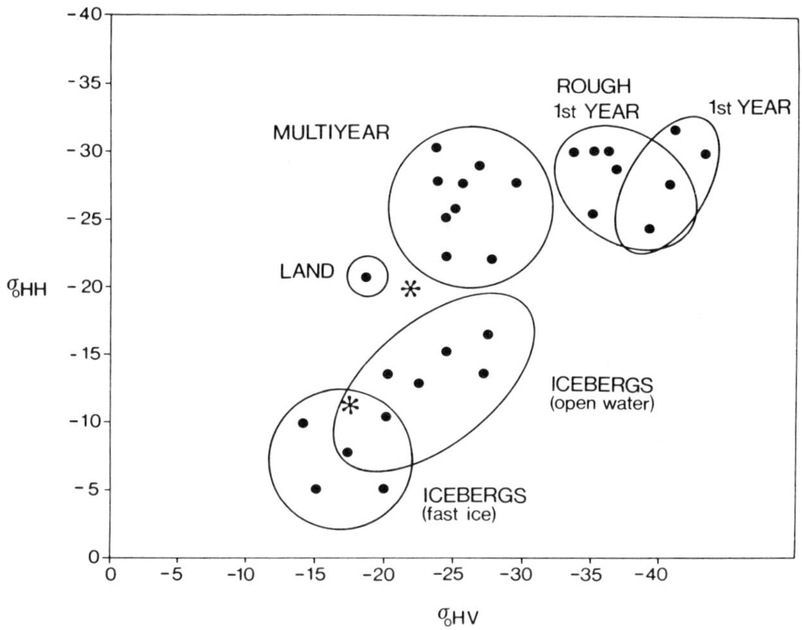

Fig. 8.26 Plot of like-polarized versus cross-polarized radar cross-sections, σ_o, for ice targets. The open water iceberg values were collected at X-band, all other values are for K_u-band.

while the other data (with the exception of the two asterisked points) were collected at K_u-band. Therefore, it is not surprising that the X-band values are lower, since the radar cross-section of the ice targets increases with frequency.

The two asterisks shown represent data points at X-band collected during the 1983 fast-ice experiment, and although these values are of poorer quality (because of antenna problems), they show that the open water values are in the appropriate range. These two values suggest that the cross-sections for the open water targets are probably not significantly lower than for the fixed ice situation.

8.3 SUMMARY

In this chapter, the polarization of the radar wave was defined. Experimental results from two major ice radar programs were presented and discussed.

Results from the fixed ice cover experiment clearly showed that by employing the cross-polarized radar returns, significant improvement in the ability to differentiate icebergs and multiyear ice from first-year ice can be achieved. Target- (iceberg or multiyear ice) to-clutter (first-year ice) improvements of up to 17 dB for icebergs and 10 dB for multiyear ice were shown. Normalized radar cross-sections at K_u for horizontal and vertical transmit and the associated cross-polarized returns were presented. Icebergs and multiyear ice were shown to strongly depolarize the incident radar energy. The plot of the like-polarized versus the cross-polarized returns provided classification of the various ice types.

Analysis of the results of the open-water experiment showed a high degree of variability in the return from seemingly stable targets. Normalized like- and cross-polarized radar cross-sections for both horizontal and vertical transmit at X-band were presented. It was shown that the ice targets are strong depolarizers of the incident radar energy. By employing both like- and cross-polarized radar returns, significant improvements in target detectability can be achieved.

CHAPTER 9
Backscattering from Ice

9.1 INTRODUCTION

A description of the mechanisms that control the interaction of a surface-transmitted radar signal and various types of ice can now be formulated with reference to the experimental results in Chapters 7 and 8 on frequency and polarization effects, and a review of current theoretical models for radar backscatter.

It is first appropriate to reiterate the pertinent properties of the ice types under consideration. First-year sea ice is a high-loss dielectric medium that strongly attenuates radar energy. As shown in Table 7.2, the skin depth for first-year ice at X-band is about 5 cm. Therefore, there will be little penetration of the radar wave into the ice and the signal returns will result mainly from surface scattering. The surface of first-year ice can be heavily deformed, producing significant vertical relief. With the high-loss, therefore reflective surface, considerable energy can be expected to return to the radar.

Multiyear sea ice has low salinity (particularly in high relief areas) and is a low-loss dielectric. Table 7.2 shows that the skin depth for multiyear ice is considerable, resulting in significant penetration of the radar wave into the ice. This penetration will permit interaction between the radar wave and the inhomogeneities in the ice. The top layer of multiyear ice contains a concentration of relatively large (1 to 2 mm) air bubbles or voids. Multiyear hummocks and ridges have significant vertical relief, thus providing large local grazing angles for the microwave energy.

Ice of glacial origin (icebergs and their fragments) is salt-free and is a very low-loss dielectric. As Table 7.2 shows, the skin depth for glacier ice is also considerable, with low attenuation of radar energy permitting deep penetration of the signal. This results in interaction between the radar wave and any inhomogeneities in the ice. The inhomogeneities usually consist of small air bubbles (0.1 to 0.5 mm) entrapped in the ice during its conversion from snow, and cracks due to pressure. Icebergs and bergy bits exhibit significant vertical relief; hence, they have appreciable local grazing angles for penetration by the radar wave. Growlers, on the other hand, have a low profile and can be well-smoothed, providing local grazing angles equal to the radar grazing angle. In open water, the detection problem is further compounded by the effect of wave wash and sea spray which wet the surface of the growler and may be expected to reduce signal penetration into the ice.

9.2 THEORETICAL STUDIES

Over the past fifteen to twenty years considerable effort has been directed toward the modelling of radar scattering phenomena. However, it is only in recent years that models have been developed that agree reasonably well with experimental results.

Early work by Fung (1967) and Beckmann and Spizzichino (1963), amongst others, concluded that polarized backscatter was related exclusively to surface statistics and that depolarization resulted from edge effects only; thus early models could not account for any significant depolarization. Leader (1970) and Rouse (1972) showed that the depolarization was not exclusively a surface phenomena but was related to a subsurface scattering effect. Rouse (1972), in fact, showed that for a single surface the depolarization power was, to a first order, completely dependent upon the volume scattering. Rouse concluded that the polarized return also contains a contribution from the subsurface and that for highly reflective surfaces, multiple scattering at the surface is enhanced and the subsurface contributions decrease.

Parashar (1974) demonstrated the effect of volumetric scattering from ice using airborne radar. This scattering model contained an expression for like-polarized cross-sections based on surface and volume correlation lengths and standard deviations, but yielded a cross-polarized cross-section of zero.

Blanchard and Rouse (1980), following up on the work of Rouse (1972), used two formulations to account for each of surface and subsurface scattering. They concluded that the like-polarized cross-section consists of contributions from both the surface and the subsurface, and that as the volume reflection coefficient increases the contribution from the subsurface begins to dominate. They also concluded that the depolarized energy was independent of surface conditions and was only a function of volumetric scattering geometry and subsurface permittivity. Sheives (1974) noted that a low density of scatterers in the volume caused only single scattering and did not contribute to depolarization, whereas a high particle density resulted in significant depolarization.

Fung and Eom (1982), using an intensity rather than a field approach with a combined rough surface and volume scattering theory, modelled backscatter from sea ice.

Their results, which agree well with observed data, suggest the following:

1) The cross-polarized return results from multiple scattering in the volume that is enhanced by the rough surface.
2) For near-vertical incidence angles, the backscatter is dominated by surface scattering.
3) Beyond 25° incidence angle, the like-polarized return is dominated by volume scattering, which in turn is controlled by the permittivity of the medium and the rough surface boundary.
4) The absolute value of the like- and depolarized-returns and their separation is controlled by the scattering albedo of the volume and the boundary roughness.

Ellis (1984), used the radiative transfer theory (accounting for volume and surface scattering) to model the scattering from icebergs. He modelled his results at X- and K_u-band, and showed that K_u-band returns should be stronger than X-band and that there is little difference between horizontal (HH) or vertical (VV) returns. He also showed that cross-polarized returns should be quite high, being 2 to 8 dB below the like polarizations. Ellis also looked at signal-to-clutter improvements in detecting icebergs in the ocean and in pack ice and concluded that cross polarization provides the best signal-to-clutter situation.

Kim _et al_. (1984) used two models for backscatter returns from sea ice. They employed a surface scattering model for first-year ice that ignores volume scattering contributions, and a combined volume and surface scattering model for multiyear ice. Using a physical optics model as an approximation to the Kirchoff model, and with an exponential correlation function, they showed results that are in good agreement with experimental data for scattering from lossy first-year ice and wet (summer) multiyear ice.

Using the radiative transfer method (intensity rather than a field solution) Kim et al. (1984) showed that the volume scattering characteristics are controlled by the dielectric constant of the ice and the size of the air bubbles. For multiyear ice, air bubbles in the recrystallized ice layer at the top were assumed to be the primary volume scatterers. They noted that surface scattering is always present and that at frequencies below X-band the volume scattering contribution for multiyear ice can be lower than the surface scattering contribution. It is concluded that depolarization is a secondary effect for the surface scattering associated with first-year ice, while volume scattering causes considerable depolarization, and this should facilitate discriminating multiyear from first-year ice.

All the modelling results discussed have assumed that any scatterers within the medium have been Rayleigh scatterers. If the scatterers begin to become non-Rayleigh, the modelling results cease to apply.

From the modelling results it is reasonable to assume that for high-loss first-year ice, the major scattering mechanism is surface scattering. As the modelling efforts discussed previously do not account for any depolarization caused by surface scattering, the depolarization that does occur must result from multiple scattering at the surface, and (possibly) from a small amount of volume scattering in the near-surface layer. The low average values of cross-polarized σ_0 discussed in Chapter 8 and the high like- to cross-polarized ratios are consistent with this explanation. The radar returns from highly saline first-year ice can be assumed to be predominantly the result of surface scattering, and as such will be related directly to surface roughness.

From the modelling results discussed, the returned energy from low-loss dielectrics such as glacier and multiyear ice will be assumed to be the result of three scattering mechanisms: surface scattering, multiple scattering at the surface, and volume scattering. The contribution of each will depend on the roughness of the surface and the size of the scatterers with respect to the wavelength within the volume.

9.3 FREQUENCY EFFECTS

Weeks and Ackley (1982) stated that if the scatterers in the volume are less than about 0.1 wavelengths of the incident radiation, then the contribution of the scatterers can be neglected. Conversely, as the particle sizes approach the wavelength of the radiation, then scattering by the individual particles becomes significant.

For typical multiyear ice with near-surface scatterers of about 2 mm, according to Weeks and Ackley (1982), the wavelength should be no longer than 2 cm for the individual scatterers to have an effect on the total return. Therefore, for multiyear ice it is anticipated that significant scattering from the inhomogeneities will not result at X-band but will result at K_u- and K_a-band. For glacier ice with bubble sizes assumed to be 0.5 mm, the radar wavelength should be no longer than 0.5 cm for scattering from the inhomogeneities to be significant, so that it is not anticipated that even K_a-band will show strong evidence of scattering from the inhomogeneities.

The situation can be appreciated in another fashion by looking at Figure 9.1. This figure (based on Skolnik, 1980), plots the radar cross-section of a sphere as a function of the ratio of circumference of the sphere to the radar wavelength. Targets that are small compared to the wavelength are Rayleigh scatterers. They will have

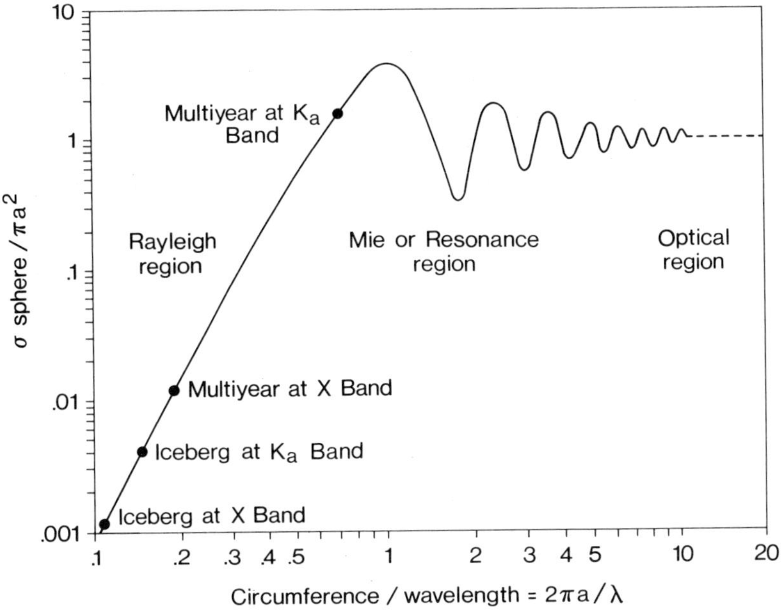

Fig. 9.1 Scattering cross-section for air bubbles in multiyear and glacier ice at X- and K_a-band (based on Skolnik, 1980).

small radar cross-sections, and therefore scatter little energy. As the size of the particle increases relative to the wavelength, the cross-section will increase, resulting in the scattering of more energy. Superimposed on this figure are points representing scatterers in multiyear and glacier ice at X- and K_a-band. The scatterers in the multiyear ice are again assumed to be 2 mm in diameter, while the scatterers in the glacier ice are assumed to be 0.5 mm in diameter. It is apparent that at the same frequency the scatterers in multiyear ice have a larger cross-section than the scatterers in glacier ice, and as the frequency increases the cross-section of the scatterers in a particular ice type increases towards the optical limit.

Kim et al. (1984) modelled the effect of air bubble size on the scattering albedo where the albedo, ω, is

$$\omega = K_S/(K_S + K_a) \tag{9.1}$$

and where the absorption coefficient, K_a, is related to the dielectric constant of the ice and the volume of air, and the scattering coefficient, K_S, is related to the size of the scatterers relative to the wavelength, the density of the scatterers, and the dielectric of the scatterers and the surrounding volume. Figure 9.2 from Kim et al. (1984) shows the albedo as a function of frequency for three bubble sizes. Again, it is clear that although the 2 mm bubbles in multiyear ice at X-band will scatter some energy, the albedo increases significantly at K_u-band (16 GHz). It is equally clear that the 0.5 mm bubbles in glacier ice have a low albedo even at K_u-band.

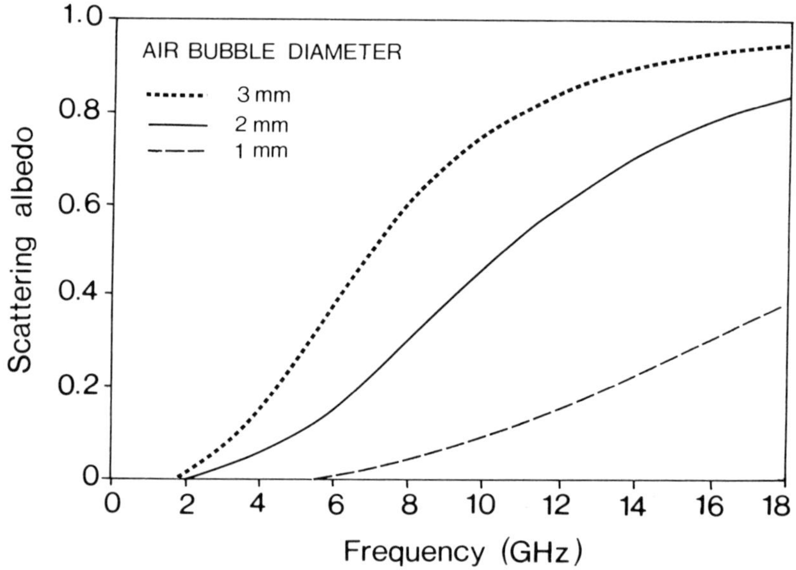

Fig. 9.2 Effect of air bubble size on scattering albedo; density = 0.7 g/cm³, $\varepsilon = 3.15 - j0.01$ (Kim et al., 1984).

The validity of these assumptions can be judged by reviewing the results in Chapters 7 and 8. As shown in Table 7.6, the average σ_0 values for first-year ice, multiyear ice, and glacier ice increased by 35 dB, 50 dB, and 20 dB, respectively, from S-band to K_a-band. Weathering of multiyear ice during the summer melt makes the surface smoother, in a large-scale sense, than the first-year ice, although it may have the same vertical relief. Assuming that the surface scattering is directly related to surface roughness, as the frequency increases from S- to K_a-band (causing the ice to appear rougher to the radar) the rougher first-year ice would be expected to show a larger increase in backscatter than the smoother multiyear ice. However, the σ_0 of multiyear ice increased 15 dB more than the first-year ice. Therefore, the increase in the σ_0 value of multiyear ice over and above the increase for first-year ice must be caused by volume scattering.

Another effect of increasing the radar frequency is to direct more radar energy onto the ice surface (as discussed in the Pattern Propagation Factor section of Chapter 6). As a result, both the first-year and multiyear ice will exhibit an additional increase in backscatter as the frequency increases because of this improved surface illumination. The iceberg, with larger vertical extent, will not benefit from this effect to the same degree. For example, the σ_0 value for glacier ice increased by only 20 dB as compared to the significantly larger increases for first-year and multiyear ice. The larger part of the 20 dB increase for glacier ice is attributed to volume scattering. However, the volume scattering effect is significantly less than in the multiyear ice because of the smaller sizes of the scatterers in the glacier ice.

Table 7.6 shows the change in σ_0 values from K_u- to K_a-band: the average first-year σ_0 increased by 16 dB, the multiyear σ_0 increased by 18 dB, and the glacier ice increased by 4 dB. The increase in σ_0 for first-year ice is caused mainly by the increase in roughness of the first-year ice at K_a-band. However, it is apparent that there may be, in fact, some volume scattering from the first-year ice as well. First-year ice contains air bubbles of about 0.5 mm (Chapter 3) as well as brine drainage pockets. As the first-year σ_0 values were collected in the late spring it is probable that some of these pockets contained air voids. Although the radar signals would not penetrate far into such saline ice, these voids near the surface could contribute to some volume scattering.

The increase in multiyear ice σ_0 value from K_u- to K_a-band (Table 7.6) was partially a result of the increased surface roughness at K_a-band but, to a greater extent resulted from increased volume scattering from the inhomogeneities in the ice. The σ_0 for glacier ice, on the other hand, has on average only increased 4 dB from K_u- to K_a-band. It is concluded that, as the surface roughness would have increased at K_a-band there could have been little contribution to scattering from the small air pockets in the glacier ice.

It is anticipated that the volume scattering for glacier and multiyear ice is made up of two components. There is scattering from surfaces within the volume as well as scattering from the air bubbles within the volume. For multiyear ice, there is substantial scattering from the air bubbles as well as a contribution from the internal surfaces. For glacier ice, the volume scattering is largely from the internal surfaces, since the air bubbles are considerably smaller. This results in considerably less contribution from volume scattering for glacier ice.

In summary, as the wavelength of the radar signal approaches the physical size of the inhomogeneities in the ice, significant increase in returned energy will be experienced because of enhanced volume scattering. For multiyear ice, this occurs at frequencies above X-band and for glacier ice it is anticipated to occur above K_a-band. The experimental results also suggest that there may be some volume scattering from the late season first-year ice.

9.4 POLARIZATION EFFECTS

As discussed in the review of theoretical modelling of scattering, surface scattering does not depolarize the incident energy. Multiple scattering at the surface results in some depolarization, although the majority of the depolarized energy is expected to result from volume scattering.

Table 8.2 in Chapter 8 shows that for K_u-band, the ratio of like- to cross-polarized σ_0 for multiyear ice is 0.4 dB on average, indicating that multiyear ice is a strong depolarizer. As indicated in Chapter 8, the K_u-band results were examined because of the low return levels from the K_a-band radar, particularly for cross-polarized signals. Table 8.2 shows that glacier ice is another reasonably strong depolarizer, with an average like- to cross-polarized σ_0 ratio of 6.5 dB. Although both ice types are penetrated easily by radar waves, the multiyear ice depolarized the radar signal by about 6 dB more than the icebergs. This larger depolarizing effect in the multiyear ice was caused by a larger contribution from volume scattering from inhomogeneities within the ice. Again referring to Table 8.2, rough first-year ice had an average like- to cross-polarized ratio of 8.3 dB, while moderately-rough first-year ice had a ratio of 13.3 dB. It is therefore proposed that much of this depolarizing effect resulted from surface roughness; that is, multiple scattering at the surface, although there may be some small contribution from volume scattering caused by air bubbles in the volume.

The relatively small difference between the depolarizing ratio for glacier ice and rough first-year ice is accounted for by the depolarization in both cases coming mostly from multiple scattering at the surface. The 2-dB higher ratio for the icebergs probably results from the multiple scattering from surfaces within the volume of the icebergs. In contrast, the strong depolarization occurring in multiyear ice is caused mainly by volume scattering.

In the analysis of the 1984 open-water data it was shown that icebergs in open water also depolarize the radar energy to a significant extent. Unfortunately, the sea also depolarizes the radar wave such that there is little or no target-to-clutter improvement in using the depolarized returns.

9.5 SUMMARY

This chapter formulated an explanation for the radar backscatter from ice which agrees with the experimental results discussed. The chapter started with a review of radar scattering theory in general, then in particular as it applies to ice. The chapter continued with an analysis of the frequency and polarization effects on the radar returns from ice, and formulated the following scattering process which is consistent with the experimental results.

The theoretical models and experimental data suggest that for first-year ice the main scattering mechanism is surface scattering, with multiple scattering at the surface contributing to and causing a small amount of depolarization. There is a possibility that there may be a small contribution from volume scattering in the late-season first-year ice.

The scattering from multiyear ice is a result of surface scattering, multiple scattering at the surface, and predominantly, volume scattering. The depolarization of the radar wave is caused to a large extent by volume scattering.

For glacier ice there is extensive penetration of the radar wave into the ice. The scattering mechanisms controlling the return of energy are surface and multiple scattering. The relatively smaller depolarization, when compared to multiyear ice, is mainly a result of multiple scattering at the surface and scattering from internal surfaces within the volume.

These conclusions are consistent with the smaller X-band depolarization noted in the 1983 data, while K_u- and K_a-band radars produced more substantial depolarizing effects. They also explain the stronger depolarizing effect in multiyear ice than glacier ice and the strong like-polarized returns from multiyear ice at K_a-band, as there are larger scatterers in multiyear ice.

CHAPTER 10
Other Radar Parameters

There are three other radar parameters that have an effect on the ability of the radar to detect and classify ice targets. These parameters are the height of the radar antenna, the azimuthal beamwidth of the antenna, and the transmitted pulse length. The latter two define the resolution of the radar.

10.1 ANTENNA HEIGHT

The height of the radar antenna above the sea surface has a major effect on the performance of the radar system. It is intuitively obvious that for longer range detection, a higher antenna is preferred. This has been the conventional wisdom and is demonstrated in shipborne installations where the radar antenna is located as high as possible. For most conditions this wisdom is well founded, although there are some exceptions.

The radar horizon, for a normally refractive atmosphere, was shown in Equation 4.10. The range increases as a function of the square root of the height of the antenna. For an antenna height of 15 m, the radar range to the surface is 8.6 nm, while for a 30-m antenna the range to the surface is 12.2 nm. The higher antenna results in a significant improvement in the maximum range of detection, given sufficient signal return.

In addition to affecting the range to the radar horizon and maximum range of detection, the antenna height affects the propagation factor, F, as Equation 6.2 shows. Referring to Figure 6.9 a higher antenna results in the first antenna lobe maximum occurring closer to the surface, putting more energy on low-lying targets and improving the probability of their detection.

As a further consequence of the multipath phenomenon described in Chapter 6, the target power return drops off as $1/R^4$ out to the transition range, and thereafter drops off at a $1/R^8$ rate. This effect results in targets beyond the transition range being illuminated with significantly less power. As Equation 6.4 shows, the transition range is directly proportional to the height of the antenna and thus by increasing the height of the antenna the transition range can be extended, which will result in improved probability of detection out to further ranges.

As shown in Figure 4.3, as the height of the antenna increases, the grazing angle for a particular range also increases. This increased grazing angle will allow increased penetration of the radar signal into the ice target; in the case of classification of ice targets, it will allow more volume scattering and will improve the ability to differentiate ice types. In the case of the detection of an ice target in open water, the increased penetration will also result in more volume scattering, and improve its detectability by separating the ice target from the sea clutter.

One possible drawback of increasing the height of the antenna is that as the height of the antenna increases, the range to which sea clutter is a problem also increases. For larger targets, the extended range of the sea clutter and its increased intensity may visually mask the returns from the target. However, to detect small targets it is necessary to have the radar energy on the surface and, as shown in Chapter 6, for a fixed radar frequency this is achieved by raising the antenna. As noted by Williams (1979), to detect small targets it is first necessary to display the clutter. Once the surface has been illuminated as evidenced by the appearance of clutter, clutter suppression techniques can be employed to make the target detectable.

The other potential difficulty with raising the antenna height is related to the ducting phenomena. The propagation ducts anticipated to be encountered in the Canadian offshore area are generally shallow, near-surface ducts, and the optimum placement of the radar antenna to take advantage of such ducting conditions is within the duct; therefore lower, near-surface placement of the antenna is suggested. However, as noted in Chapter 6, although the optimum placement of the antenna is within the duct, antennas placed above the duct experience extended propagation ranges because of the leaky nature of the duct, although not to the same extent as a lower antenna.

Several points related to antenna height are illustrated in Figures 10.1 and 10.2. These figures show images of radar returns from a full ice cover with an X-band radar, collected during the 1981 field experiment. The radar conditions for both images are identical, except that Figure 10.1 is for the radar antenna mounted at 37 m, while Figure 10.2 is for the antenna mounted at 29 m. Although the difference in elevation between the two antennas is minimal, there are significant differences between the two images. Figure 10.1, from the higher antenna, has stronger signal returns and more texture, particularly at far range, resulting from more energy reaching the ice because of the multipath factor. Figure 10.1 also shows detection of targets at greater ranges, particularly in the area of A.

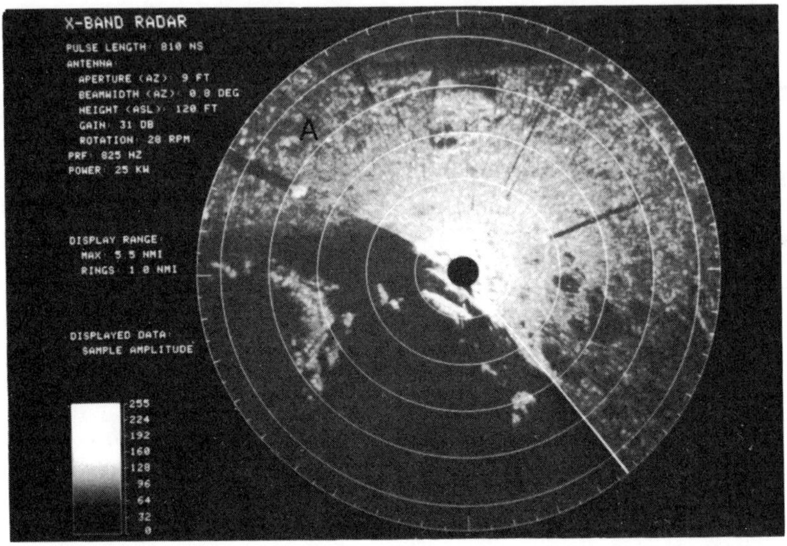

Fig. 10.1 X-band image for radar antenna mounted at 37 m.

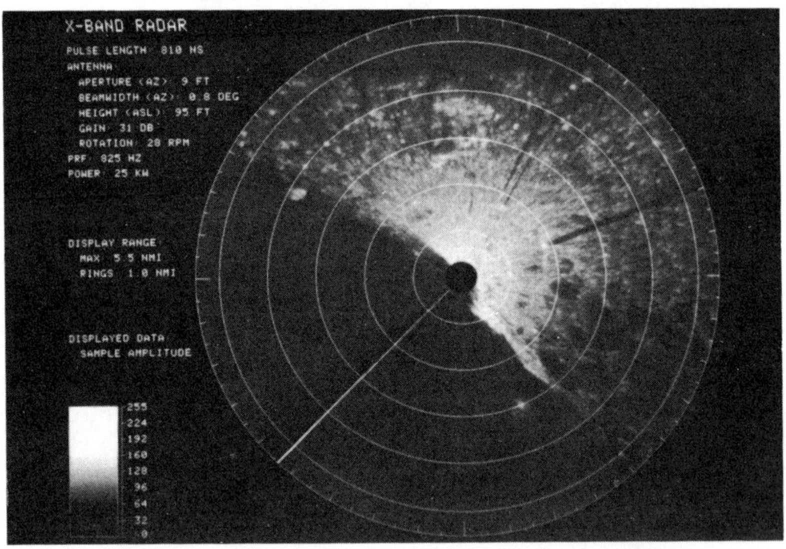

Fig. 10.2 X-band image for radar antenna mounted at 29 m.

Figure 10.2 shows longer shadows behind the large targets as a result of the lower height of the antenna. In addition, the image shows more discrete targets in the intermediate ranges, because there is less energy near the surface, caused by the multipath factor, and only targets with some significant vertical extent can be illuminated. These effects lead to a reduction in the indication of area roughness.

10.2 RADAR RESOLUTION

The radar resolution is defined as the ability to separate two targets. It has both a range and an azimuthal component. The resolution in range is determined by the pulse length of the radar while the azimuthal resolution is determined by the beamwidth (3 dB) of the antenna. The effects of varying the pulse length and the azimuthal resolution are examined separately below.

10.3 AZIMUTHAL BEAMWIDTH

The azimuthal beamwidth of the antenna, as described in Chapter 4, controls the azimuthal resolution (Equation 4.14) of the radar.

As the antenna beamwidth increases from 1 to 2°, the resolution is halved; that is, for two targets to be resolved independently they must be twice as far apart in bearing with a 2° antenna as with a 1° antenna. As the spatial azimuthal resolution is also a function of range, the spatial resolution in azimuth for a wide beamwidth antenna can become very large as the range increases.

In the target detection mode in sea clutter, the wide beamwidth antenna results in large clutter cells. The illuminated area is equal to the azimuthal resolution times the range resolution, so that a doubling of the azimuthal beamwidth results in a doubling of the clutter cell, and subsequently causes a reduction in the target-to-clutter signal ratio and a reduced detection probability. However, both Croney *et al.* (1975) and Williams (1975) show examples of reduced target detection with higher azimuthal beamwidth antennas. The explanation of this is two-fold. First, the wider azimuthal beamwidth antenna smears or broadens the return from the target and thus makes it more recognizable. Second, as discussed earlier in Chapter 6, as the resolution of the radar system is increased, the statistics of the returns depart from Rayleigh and develop a higher probability of large returns. It is these large returns that make the target more difficult to see.

As just noted, the wider-beamwidth antenna results in a smearing of target returns in azimuth. In many cases, as in the detection of a ship at sea, this smearing results in easier identification of the target. However, for ice classification this can be a disadvantage, as shown by Figures 10.3 and 10.4. The figures show images of the radar returns with an X-band radar for a full ice cover, collected during the 1981 experimental program. The radar parameters for the two images are identical, except that the image in Figure 10.3 was collected with a 2.8-m antenna with an azimuthal beamwidth of 0.8°, while the image in Figure 10.4 was collected with a 1.2-m antenna with an azimuthal beamwidth of 1.9°. Figure 10.3 is a sharper image, while Figure 10.4 is blurred or smeared. In particular, the shadows behind the large ice targets at A and B were completely filled in by the larger beamwidth in Figure 10.4, thus significantly decreasing their visibility. The shadow behind target C was also reduced, but was still visible. Shadows are a valuable indication that a target has significant vertical relief. Target C was less visible because of the smearing of its return, as were other targets, particularly at intermediate and far ranges.

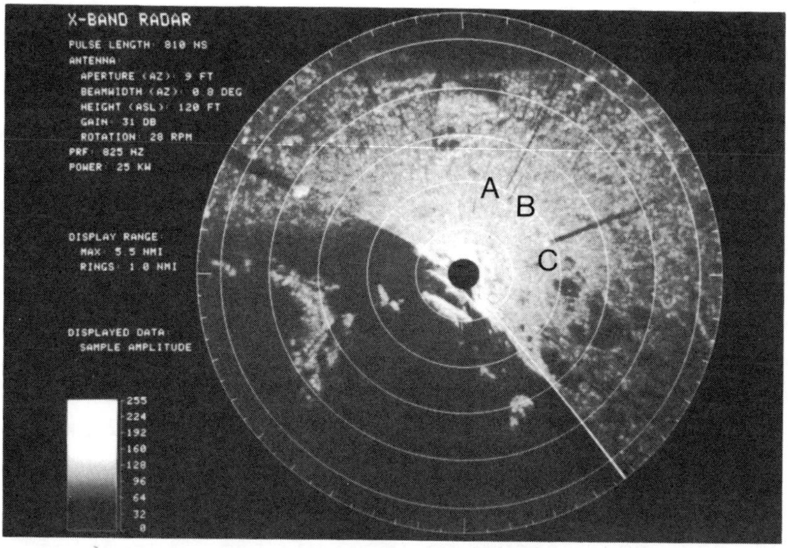

Fig. 10.3 X-band image for 0.8° beamwidth antenna.

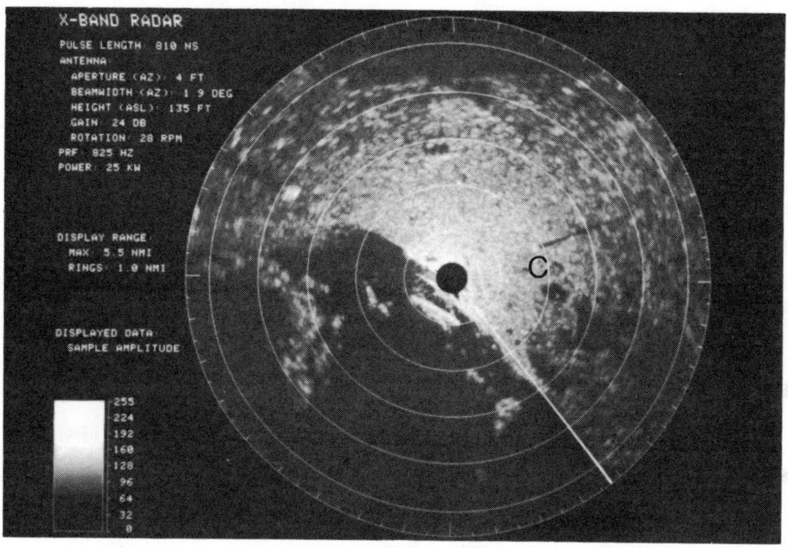

Fig. 10.4 X-band image for 1.9° beamwidth antenna.

10.4 TRANSMITTED PULSE LENGTH

The resolution of the radar in range was defined in Equation 4.12, where the resolution in range is directly proportional to the pulse length. If the pulse length is doubled, the distance in range between two targets must also double for them to be resolved individually.

Although it is desirable to have the maximum resolution possible, there is a serious problem that develops as the pulse length is shortened. The energy reflected from a target is related directly to the average power incident upon it, and since the average power is determined by the peak power times the duration of the pulse, shortening the pulse reduces the energy returned from the target. Figure 10.5 shows images from a radar with the same radar parameters, with the exception of pulse length. As the pulse length decreases from 810, to 220, to 70 ns, the image becomes sharper because of the increasing range resolution. However, also noticeable in Figure 10.5 is the decrease in range of coverage resulting from the reduction in average power. In the 70-ns image, targets at far range are lost completely, while the returns from closer targets are significantly reduced.

This problem of reduced coverage with increased range resolution can be overcome through the use of pulse compression techniques. Pulse compression uses long coded transmission pulses which, upon reception, are decoded to yield an effective pulse length much shorter than that of the transmitted pulse. The longer transmitted pulse yields higher average transmitter power and thus longer range of coverage. Compression ratios, that is the ratio of the length of the actual transmitted pulse to the length of the effective compressed pulse, are typically in the range of 50 to 200. For a compression ratio of 50, for example, a real pulse of 1 μs, with resulting

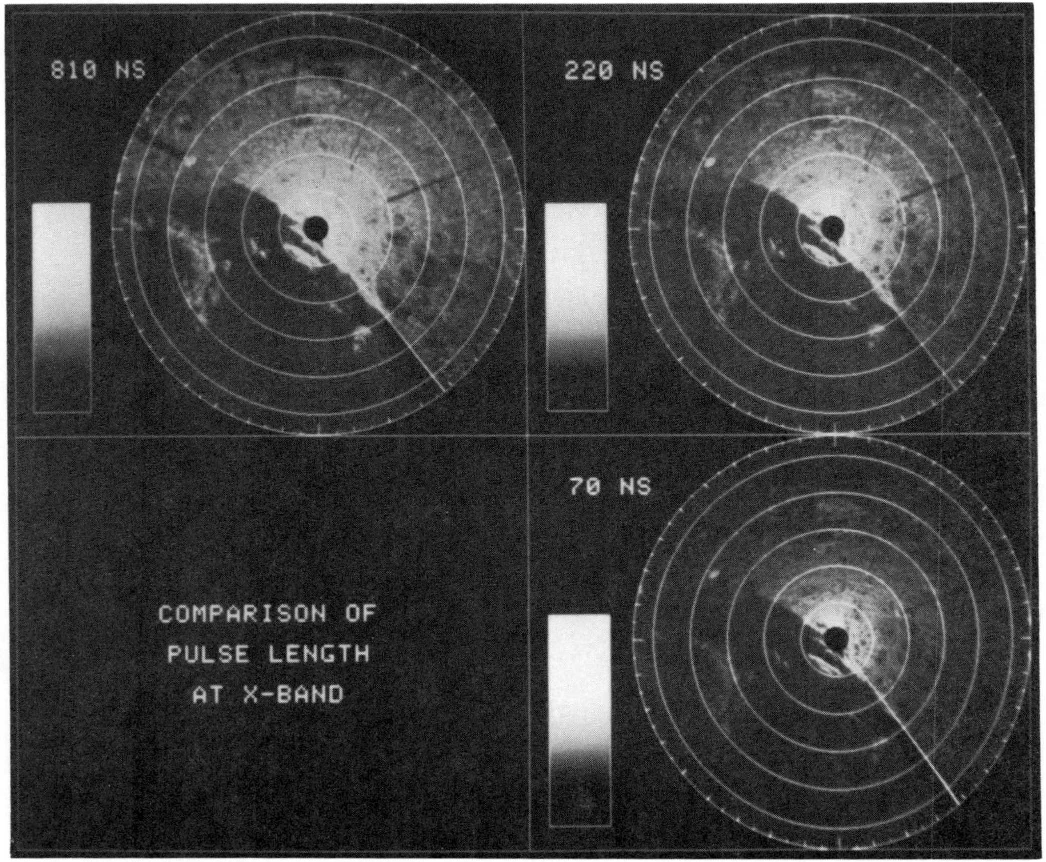

Fig. 10.5 X-band images for various transmit pulse lengths.

high average power could be transmitted, but range resolution corresponding to a 20-ns pulse could be achieved.

A further difficulty with short pulses, either natural or compressed, is their effect on sea clutter characteristics as the pulses become progressively shorter. The statistics of sea clutter returns for normal long-pulse radars (as shown in Chapter 6) have been represented historically by a Rayleigh distribution. Automatic detection algorithms and constant false alarm rate (CFAR) circuitry have been designed based on an assumed Rayleigh distribution. However, as the pulse length is shortened, the sea clutter returns become more spiky and more non-Rayleigh in nature. Further, as the sea state increases, the statistics of the clutter develop long tails with higher probabilities of strong returns. Detection algorithms and conventional CFAR circuitry operating in this environment will yield unacceptably large probabilities of false alarm. To reduce the probability of false alarms, techniques which account for the change in statistics must be employed.

10.5 SUMMARY

This chapter looked at the effects of radar antenna height, antenna azimuthal beamwidth, and transmitted pulse length on the ability to detect and classify ice. It was shown that raising the antenna height puts more energy on the ice surface and thus provides more surface detail, while at the same time the radar shadows behind targets with significant vertical relief are reduced making them less easily detected. Reducing the azimuthal beamwidth of the antenna, and thus increasing the azimuthal resolution, improves the ability to detect individual targets. Finally, by reducing the length of the transmitted pulse, the range resolution can be improved but at the expense of reduced range coverage.

CHAPTER 11
Techniques for Improving Target Detection

11.1 INTRODUCTION

One of the most severe problems remaining in navigating in ice-infested waters is the reliable detection of small ice targets (primarily growlers) in sea clutter. Williams (1979) comments that

> "in order to detect growlers, or indeed any really small targets, (a radar) must first display sea clutter out to the desired detection range if there is to be any chance at all of making a detection."

However, existing marine radars are inadequately designed for small target detection in a clutter-limited situation. Significant improvements in the reliability of detection are necessary to detect targets of interest. This chapter will review several techniques that can improve the reliability of detection.

For a non-coherent radar system using returned signal amplitude (power) only, the principles involved in the detection process can be demonstrated as shown in Figure 11.1. The interference (which can be composed of clutter and/or noise) and the signal-plus-interference are both represented by distributions of their returned powers with the mean powers as shown. The difference between the means (in dB) is the signal-to-interference ratio. For a given distribution (i.e., Rayleigh) of the

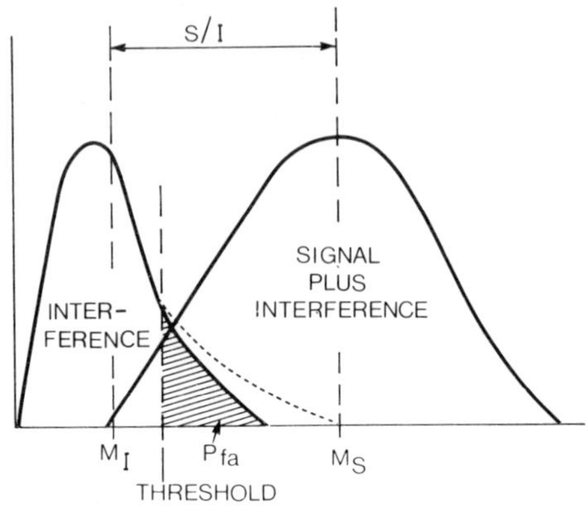

Fig. 11.1 Distributions of interference and signal-plus-interference. Probability of false alarm (P_{fa}), as shown by the shaded area, is the area under the interference curve above the detection threshold. The dotted line shows increased P_{fa} resulting from an interference distribution with a longer "tail." Probability of detection (P_d) is the area under the signal-plus-interference curve above the detection threshold. M_I and M_S indicate the means of the two distributions. The difference between the means is defined as the signal-to-interference (S/I) ratio.

interference, a threshold can be set to yield a required theoretical probability of false alarm (P_{fa}). Then with a given signal-to-interference ratio (S/I) the probability of detection (P_d) can be determined. The P_d is the area under the signal-plus-interference curve above the threshold. If the width of the interference distribution was widened as shown in the figure by lengthening the tail of the distribution, the threshold would have to be increased to maintain the same P_{fa}, and the P_d would be reduced.

Inspection of Figure 11.1 indicates two ways to improve target detection. First, target detection can be improved by increasing the signal-to-interference ratio, thereby increasing the separation of the target and interference distributions. This improvement can only be achieved by modifying the parameters of the radar. Second, detection improvements can be achieved through signal processing by reducing the widths of the distributions of the interference and the signal, hence their overlap, so that the same P_{fa} and P_d can be maintained for a smaller signal-to-interference ratio.

11.2 IMPROVING THE SIGNAL-TO-INTERFERENCE RATIO

The signal-to-interference ratio is the separation (in dB) between the mean power of the signal and the mean power of the interference. As Figure 11.2 (which has the same distributions as Figure 11.1) shows, if the S/I ratio is increased, the distributions of the signal and the interference move apart. If the threshold is held

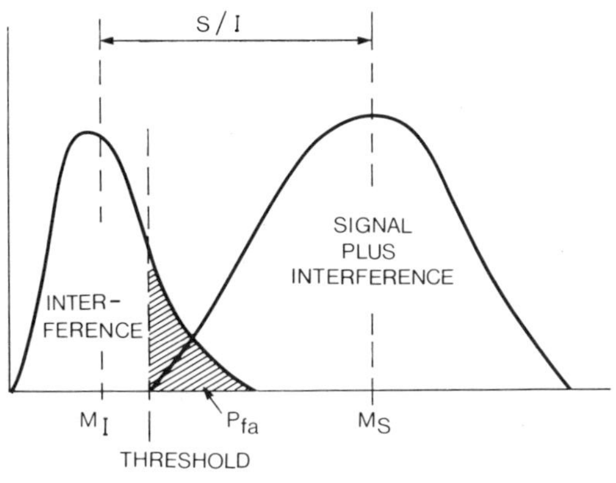

Fig. 11.2 Distributions of interference and signal-plus-interference for improved S/I ratio, hence improved P_d resulting from an increase in the mean signal (M_S) with P_{fa} held constant.

constant to maintain a given P_{fa}, then the P_d will improve.

Since the signal-to-interference ratio is the ratio between the signal power and the interference power, it is obvious that this ratio can be improved by either increasing the signal power or decreasing the interference power. Chapter 7 showed improvements in target returns at S-band and K_a-band, although the improved signal-to-clutter (interference) ratio at S-band may have been caused by reduction in the clutter return as well. In Chapter 8, substantial improvements in signal-to-clutter ratio were achieved by employing the cross-polarized return where the returns from targets of interest (icebergs and multiyear floes) were only slightly reduced, whereas the returns from the clutter (first-year ice) were substantially reduced.

The clutter power as given in Equation 4.8 can be lessened by reducing the size of the clutter cell. The two independent variables that define the clutter cell size are the pulse length of the transmitted signal, τ, and the azimuthal beamwidth of the antenna, θ. If the clutter cell size is reduced, the returned power from the target will not be reduced (since it is reasonable to assume that the size of the target is small compared to the resolution cell), while the power returned from the clutter will be reduced by the decrease in the clutter cell size. There are, however, some practical limitations to the reduction in the resolution cell size. For a low resolution system, the radar returns in general come from a large number of scatterers which contribute relatively equally to the total return power. As the resolution cell is reduced the returned power is contributed from fewer and fewer scatterers, and there is an increasing occurrence in the number of large signal values. This results in a change in the distribution of the clutter power in which the upper tail of the distribution increases. This change in the distribution increases the difficulty of determining the proper threshold in order to deliver a required P_{fa} and P_d.

The improvements in S/I ratio discussed above will all be achieved by adjusting the parameters of the radar to optimize the signal return or minimize the interference return. If, however, the radar parameters are fixed, further improvements in detection can be achieved as discussed below.

11.3 IMPROVING THE RELIABLE SEPARATION OF THE TARGET AND INTERFERENCE DISTRIBUTIONS

In contrast to the section above where improved detection was achieved by increasing the difference between the mean power of the signal return and the mean power of the interference, this section will look at ways of improving the target detectability by reducing the width, hence overlap, of the distributions of the target and the interference returns.

This improvement can be visualized as shown in Figure 11.3 where the upper diagram shows distributions for a signal-plus-interference and interference with a given S/I ratio. The means of the signal-plus-interference and interference are shown, as well as a threshold. With this given threshold, the P_{fa} is as shown, while the probability of detection (P_d) is the area under the signal-plus-interference curve above the threshold. When the distributions of the two curves are narrowed, as shown in the lower diagram, the threshold can be lowered while maintaining the same P_{fa}, and the probability of detection will be increased.

A more conventional way to view this improvement in detection by narrowing of the distributions is shown in Figure 11.4. Again the upper diagram shows the

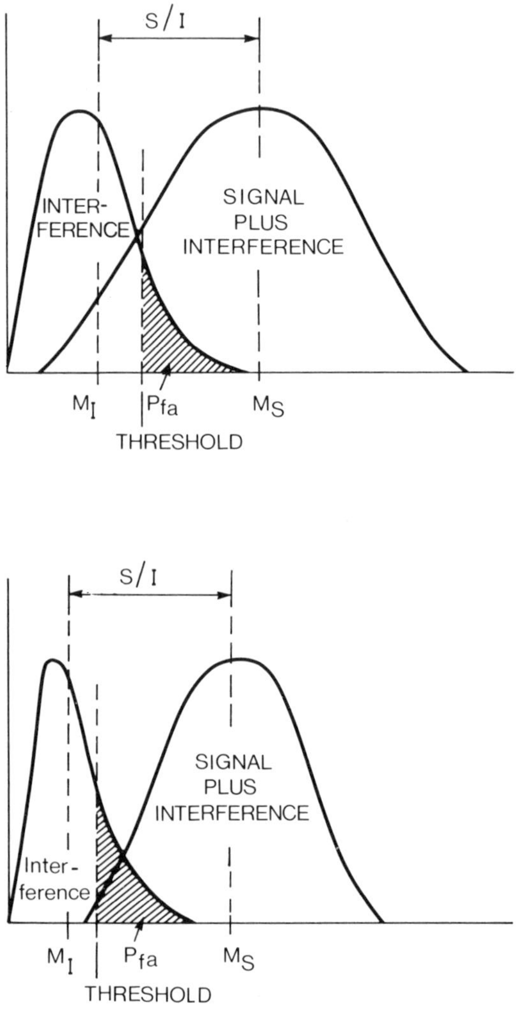

Fig. 11.3 Improved P_d for a given S/I ratio resulting from a narrowing of the distributions. Threshold has been lowered to maintain constant P_{fa}.

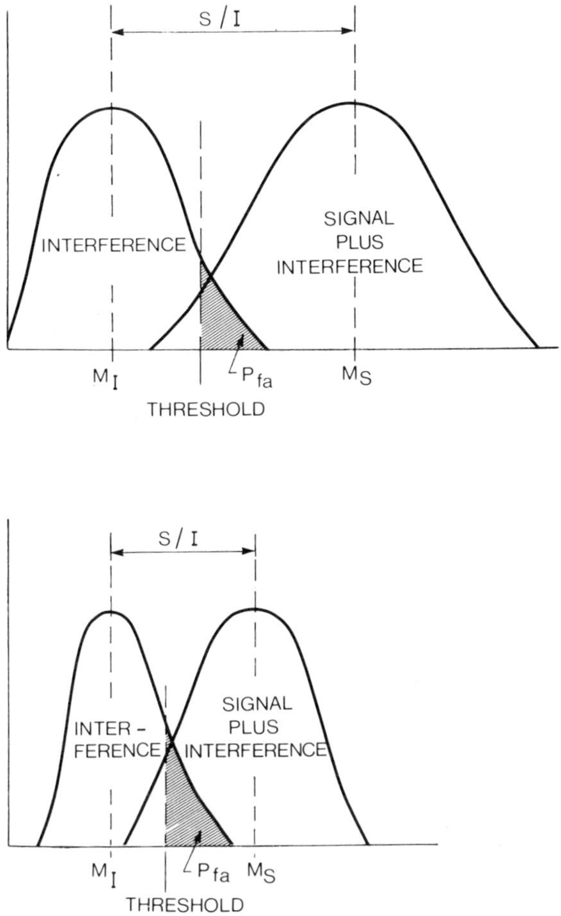

Fig. 11.4 Reduced S/I ratio required to maintain fixed P_d and P_{fa}, with reduced width of distributions. With a narrowing of the distributions the same P_d and P_{fa} performance can be achieved for a lower S/I ratio; that is, for a weaker signal.

distributions for the signal-plus-interference and interference, with the threshold set to yield a given P_{fa} and P_d for the specified S/I ratio. However, if, as shown below, the distributions can be narrowed, the same P_{fa} and P_d can be achieved with a considerably lower S/I ratio. This reduction in the required S/I ratio is often viewed as a gain in the original S/I ratio.

For a non-coherent radar, the narrowing of the signal-plus-interference and interference distributions is achieved primarily by integration.

Integration

In a non-coherent radar system using a PPI display with long-persistence phosphor, successive returns from a target tend to integrate or add up, whereas uncorrelated noise pulses or clutter returns do not. This integration results in an improved ability of an observer to detect the target.

If the successive returns from the target and the interference are retained electronically, an improvement in target detectability (as shown in Figure 11.3) can be realized. It is generally recognized that by integrating N pulses, the required signal-to-interference ratio for a given P_d and P_{fa} is reduced by a factor between $\sqrt{N}$ and N dB (Skolnik, 1980). However, to achieve the improvements offered by integration, the successive returns must be uncorrelated.

For the case of noise, samples separated by at least the radar pulse length are uncorrelated, which means that successive samples in range or samples on a pulse-to-pulse basis can be integrated to achieve signal-to-noise improvements, as shown in Figure 11.3. Figure 11.5 from Blake (1969) shows the reduction in required S/N ratio achieved by the integration of a number of pulses for a Swerling Case 1 Fluctuating Target for $P_d = 0.50$.

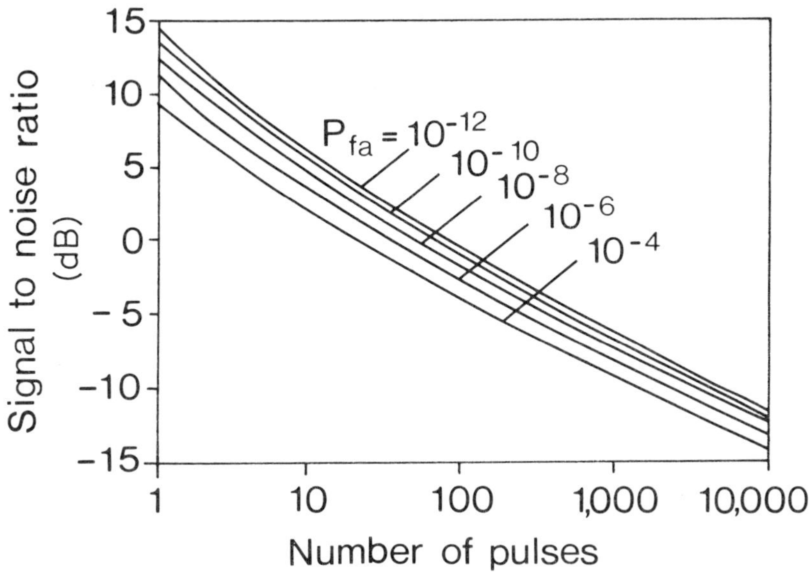

Fig. 11.5 Required signal-to-noise ratio for detection with non-coherent integration of pulses: square-law detection, Swerling Case 1 fluctuation, P_d = .50 (Blake, 1969).

As indicated in Chapter 6, with the exception of long-duration sea spikes which may take several seconds to decorrelate, sea clutter in general decorrelates in about 10 ms. For a conventional X-band marine radar with 0.8° azimuthal beamwidth, pulse repetition interval of 1 KHz, and antenna rotation time of 2 s per revolution, a target will be illuminated for about four successive pulses as the antenna scans past. The four samples will be 1 ms apart, which means that successive pulses will not yield decorrelated samples of sea clutter suitable for signal-to-clutter improvements by integration. However, pulses from successive scans (about 2 s apart) will be uncorrelated. The integration of returns from successive scans is known as scan-to-scan integration and provides substantial improvements in the signal-to-clutter ratio for the sea clutter situation.

As the target and the sensor are likely to be moving, and returns can only be integrated for the period during which the target remains within a given resolution cell, there is a limit to the number of scans which can be integrated. Two techniques which provide more uncorrelated samples within a fixed period are the use of frequency agility and high-speed scanners.

A frequency agile system is capable of changing the frequency of the transmitted signal on a pulse-to-pulse basis. If the radar frequency is changed by more than the inverse of the pulse length, the returns on a pulse-to-pulse basis will be uncorrelated. This, however, is only true for the short-duration sea clutter. The longer-duration sea spikes will still remain correlated.

The second method of providing more uncorrelated samples in the same period of time is to speed up the rotational rate of the antenna so that the time between scans is reduced. Croney (1966) described an experiment in which the radar antenna was rotated at 1,000 rpm. The high scan rate provided about 6 dB improvement in signal-to-clutter visibility for small targets in sea clutter. Croney noted, however, that to avoid severe losses in sensitivity at least two pulses must be transmitted for every beamwidth. Smith (1984) described the implementation of fast scan (150 rpm) processing in the APS/134 airborne surveillance radar system to achieve improved detection of submarine periscopes.

The actual integration of the uncorrelated signals can be accomplished in a number of ways, two of which are described below. The analogue output of the detector can be summed over a number of echoes and then applied to a detector. An integrated return exceeding the threshold will be declared a target. Conversely, a threshold can be applied to each uncorrelated return and then a final decision made based on how many times the threshold was exceeded out of the total opportunities (M of N detection).

11.4 CONSTANT FALSE ALARM RATE PROCESSORS

The preceding discussion was based on the premise that the interference power is stationary; that is, it does not vary with range or azimuth. This is, in fact, the case for the noise power; however, the clutter power varies with both range and azimuth.

If the clutter power mean is not stationary but changes with range and azimuth, then the specification of the threshold by some relationship to the mean will result in wild fluctuations in the false alarm rate. This problem can be overcome by employing a constant false alarm rate (CFAR) processor which uses an adaptive threshold based on an estimate of the mean interference power at each location to maintain a constant false alarm rate. By using a number of samples around a target cell, the processor can determine a better estimate of the interference power for that location and use this estimate to establish an adaptive threshold that maintains a constant false alarm rate.

Cell-Averaging CFAR Processor

A conventional, cell-averaging (CA) CFAR processor is shown in Figure 11.6. The processor consists of a window that slides along in range together with an algorithm that calculates a detection threshold. The samples in the window are assumed to be statistically independent, random variables, and the clutter is assumed to be uniform in the sampled area. The window samples cells on either side of the reference cell to determine an estimate of the mean interference power. The sample values are taken from a tapped-delay line with tap-spacing approximately equal to the radar pulsewidth. Because the

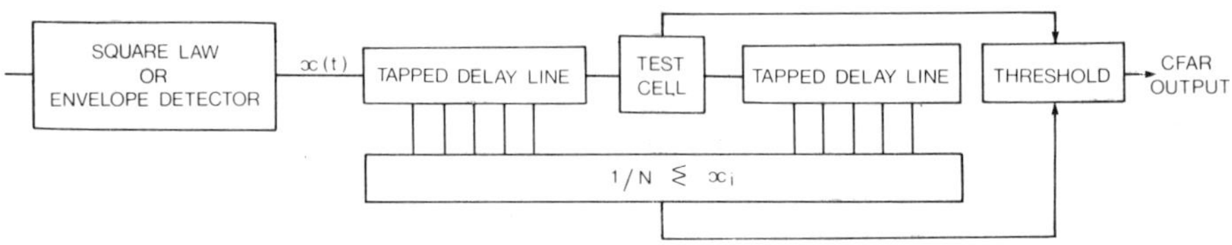

Fig. 11.6 Conventional constant false alarm rate (CFAR) processor.

signal power can spill over into adjacent cells and affect the estimate of clutter power, guard cells are established on either side of the reference cell. In the CA CFAR processor the estimate of the clutter power is derived by taking an arithmetic mean of the power in the adjacent cells. This estimate is then multiplied by a scaling factor that depends on the estimation method and the desired false alarm rate. The result of this operation is used as the threshold and is compared against the reference cell to declare, or not to declare, a target.

When the noise is a stationary Gaussian process, the CFAR processor experiences a degradation in detection performance known as CFAR loss. The CFAR loss is the increase in S/N ratio required to maintain detection performance. Figure 11.7 shows a plot of CFAR loss in dB versus the ratio of the false alarm exponent to the window size, for various signal-to-noise ratios. For a probability of false alarm of 10^{-6} and relatively large window sizes (N = 16 to 20) the CFAR loss would be less than 2 dB.

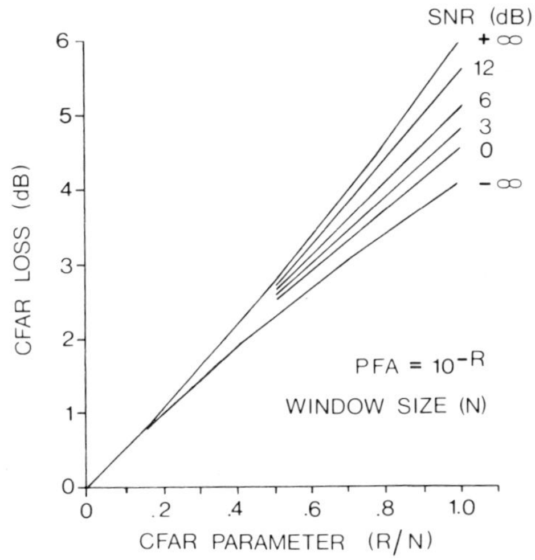

Fig. 11.7 Cell-averaging CFAR loss. Comparison of two CFAR methods used with square law detection of Swerling I targets. Moore and Lawrence (1980), the record of the 1980 International Radar Conference, IEEE publication 80 CH1493-6 AES. Copyright © 1981 IEEE.

Cell-Averaging Greatest of CFAR Processor

If the clutter is not uniform within the CFAR window, that is, if a clutter edge is present, the CA CFAR processor will give a poor estimate of the clutter power in the reference cell and result in a significant increase in false alarms. According to Moore and Lawrence (1980), when clutter occupies half the cells of a CA CFAR processor the probability of false alarm rate increases from 10^{-6} to 10^{-3}. One method of improving the performance of the CFAR processor is to use two windows, one ahead and one behind the cell under test. A separate estimation of the mean clutter is performed in each cell and the greater of the means is chosen. This method is known as the cell-averaging greatest of, or CAGO, CFAR processor. This processor results in a reduced probability of false alarms and, as noted by Rohling (1983), "the response of the (CAGO) CFAR to the jump in clutter power is distinctly steeper than for the CA CFAR although the difference vanishes with increasing distance from the edge."

Figure 11.8 shows the superior performance of the CAGO processor compared to the CA processor when a clutter edge is encountered. The dotted line is the false alarm performance for the CA processor while the solid line is for the CAGO processor. Moore and Lawrence (1980) noted that where the clutter occupied half the cells for a CAGO CFAR processor the probability of false alarm rate only increased from 10^{-6} to 1.75×10^{-5}. The CAGO CFAR processor had only a slightly greater (about 0.3 dB) CFAR loss than CA CFAR for stationary clutter.

The performance of both CA and CAGO CFAR processors is affected by the length of the window. Small window sizes

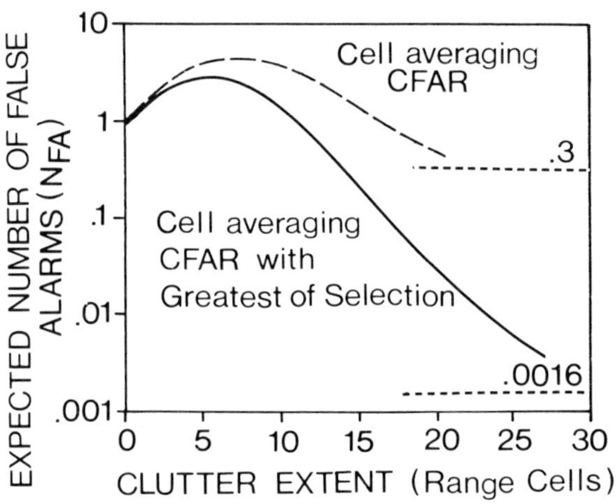

Fig. 11.8 Expected number of false alarms generated by a block of clutter for a conventional cell-averaging CFAR and for a cell-averaging CFAR with "greatest of" selection (Hansen, 1973). Reprinted by permission of the Institute of Electrical Engineers from IEE Conference Publication 105 - the IEE Radar 73, Radar Present and Future.

provide a poor estimate of the clutter power while large windows respond slowly to clutter changes (edges). The choice of the window size (N) becomes a compromise.

Other CFAR Processors

There are a number of other more sophisticated CFAR processors for specific applications. Rohling (1983) described an ordered statistics CFAR processor that quickly responds to clutter edges and is capable of detecting two targets within the clutter window. Yamada et al. (1984) and Sekine et al. (1980) described CFAR processors for operation in Weibull clutter.

11.5 SUMMARY

This chapter has shown that substantial improvements in target detection in clutter or noise can be achieved by combining the methods discussed in this chapter. Oi et al. (1984) showed an example of the detection of a ship in heavy sea clutter through the use of a system combining a CFAR processor, a pulse-to-pulse integrator, and a scan-to-scan integrator.

The detection improvements discussed in this chapter were all achieved by employing only the amplitude of the returned signal. The improvements in detection were achieved by exploiting the differences in response of the signal and the noise or clutter. These improvements can also be achieved by employing other signal spaces such as frequency, Doppler, or polarization.

Results in Chapter 7 showed improvements in target differentiation and detection in frequency space. It was shown that S-band was most suitable for detection of physically large targets, while K_a-band seemed to provide improved detection of multiyear ice.

In Chapter 8 improvements in target differentiation and detection were shown to be possible because of polarization differences between the signal and clutter. Significant improvements in the detection of multiyear and glacier ice were shown through the use of the cross-polarized channel. If a fully coherent radar was used, the full scattering matrix could be developed. It is likely that the signal and the interference returns would generate different matrix values and thus permit improved detection if the appropriate matrix locations could be specified.

With a coherent system it may also be possible to separate the target and competing signals through Doppler processing. The hope is that the sea clutter and the ice target will have separate and distinguishable Doppler frequency spectra. The Doppler frequency spectrum relates to the velocity of the target with respect to the radar. For larger ice targets (icebergs and bergy bits), the ice targets are driven by different mechanisms than the surface clutter of the sea; thus the ice and sea will have different velocities and, therefore, different Doppler spectra. However, for small ice targets (growlers), the ice moves and is driven by the same mechanisms as the sea and, therefore, at the same velocity. This will result in the Doppler spectrum of the ice target overlaying the spectrum of the sea clutter. Haykin and Currie (1984), however, have speculated that the ice target will sufficiently distort the spectrum, as compared to the spectrum for sea clutter only, such that the presence of the target will still be detected.

CHAPTER 12
Display Improvements and Image Processing

12.1 DISPLAY DEVICES

As outlined in Chapter 1, the plan position indicator (PPI) displays of marine radar systems, although inexpensive and reliable, have several severe limitations when applied to ice navigation. First, the PPI displays have a limited dynamic range of signal intensities (about 12 dB), whereas the dynamic range of ice targets may exceed 40 dB. This limitation results in either a high percentage of medium-to-large targets becoming saturated and displayed at the same intensity so that smaller targets can be displayed and detected, or the intensity being reduced to display the larger targets, thus losing the smaller, though potentially significant, targets.

Second, as a result of the spherical spreading of the radar energy with range, targets near the radar are illuminated with, and hence reflect, more energy, and are displayed more intensely than similarly-sized targets further away. Again, as in the case of dynamic range discussed, the intensity must be increased to display targets further from the radar, thus causing closer targets to saturate the display and cause "blooming."

The image on the PPI is maintained by employing tube persistence. This results in limited display intensity, and the display cannot be viewed in direct light. In daylight hours the display must be observed through a viewing hood, which restricts its operation to one person at a time. Finally, to make a range measurement on a target, the variable range marker must be made to pass through the target, which can take several scans of the radar sweep to accomplish, thus limiting the speed at which range measurements can be made.

One way to overcome these problems is through the use of a raster scan (TV-type) display. Because of its rapid update rate (1/60 of a second), the display appears to be free of flicker. The entire image is displayed continuously and cursor position measurements can be made at any time. In addition, because of the much brighter display, the device can be viewed under normal light. Most important, however, is the much wider dynamic range of this device. The wider dynamic range and use of colour allows full advantage of image processing to be taken. By using a black-and-white display with 8 bits per pixel resolution, a dynamic range of 48 dB can be accommodated, although the actual discernible dynamic range is expected to be in the 20- to 30-dB range. By using a colour display a still wider dynamic range can be accommodated, and signals in particular ranges of intensity can be enhanced by the use of colour.

A major pre-condition to employing this type of display technology is a requirement to digitize the incoming radar signal. With modern digital circuitry the analogue-to-digital conversion is readily accomplished. To accurately represent the analogue signal in digital format, it is necessary to sample and digitize the signal at least twice per range resolution cell. The range resolution is the minimum range separation for two targets to be resolved individually, and is determined by the pulse length of the transmitted radar signal. For a typical marine radar with a pulse length of 200 ns (30-m resolution), the signal must be digitized at about 10 MHz.

12.2 SCAN CONVERSION

A major consideration in using a raster scan display is that the incoming radar data is sampled in a range-bearing (polar) format, whereas the raster scan system is in an X-Y (rectangular) format. The required polar-to-rectangular mapping is known as "scan conversion." Generally, there are more sampled data points in the polar image than can be displayed in the X-Y system and this is particularly true near the centre of the image. The resulting reduction of data points in converting to the X-Y display must be managed carefully to avoid loss of information. If not corrected, the scan-conversion routine will retain the final range-bearing value appropriate for a particular X-Y co-ordinate which can easily result in the loss of a point target and could have serious implications in the case of detection of ice targets in open water. To overcome this problem peak detect-and-hold circuitry could be implemented which would assign the highest value detected to the X-Y co-ordinate, or some averaged or median value could be used.

Conversely, near the outside of the image there will be few polar data points because of the diverging nature of the polar scan lines. In fact, some X-Y locations (pixels) will not have a data value assigned. This problem results in the formation of moiré patterns at the 45° points of each quadrant of the display. These patterns are the result of missed pixels all being displayed at the same intensity.

As the location of these pixels can be determined for a given radar geometry, one solution would be to fill them with random, but relatively low-level, values. Some commercial scan conversion routines duplicate the portion of the sweep data near the display edge to fill the missed pixels. Conversely, an interpolation between two adjacent X-Y pixel values could fill the missing pixels. There are a number of scan-conversion procedures that address these and other problems. It is necessary to select the procedure that is most suitable to the image being converted, and to the nature of the targets that are to be displayed.

12.3 IMAGE PROCESSING

Range Normalization

Once the radar signal has been digitized, the last of the major problems associated with the standard PPI display can be resolved. By normalizing the radar returns as a function of range, the blooming problem described earlier can be eliminated. All targets of equal radar cross-section will then be displayed with the same intensity, regardless of their range from the radar. The normalizing process is carried out by multiplying the return signal by an appropriate factor to compensate for the range dependence. Although this normalization can be achieved with some radars by employing the sensitivity time control (STC) function, the resulting signals are uncalibrated, thus making further processing difficult.

Restoration and Enhancement

The image processing problem is approached by either restoration or enhancement of the returned radar signal. According to Andrews (1974), image restoration is defined as "reconstruction of an image towards an original by inversion of some degradation phenomena." To accomplish restoration, some knowledge concerning the degradation phenomena is required and then, through the application of analytical or statistical models, restoration of the image can take place.

Enhancement, again according to Andrews (1974), "is an attempt to improve the appearance of an image for human viewing or subsequent machine processing." Enhancement includes any technique that extracts additional information from the image that was not originally apparent. Enhancement is therefore dependent upon the image to be processed and what feature, or features, of the image are to be "improved." Additionally, the "improvement" is dependent upon what the application is and what the user is hoping to detect. Enhancement, therefore, becomes subjective, and it is difficult to prove that one image is better than another. Enhancement can include any or all of the following techniques: smoothing, high- or low-pass filtering, histogram operations, edge sharpening, bit slicing, and colour coding.

With the exception of noise reduction, which might be considered a restoration process, all processing of the ice radar images in this work can be considered as image enhancement.

Noise Reduction

Most noise in the radar image is introduced by two sources: noise (both thermal and other) associated with the radar system, and noise associated with the digitizing process. Both noise sources are white and additive, and they are usually a single sample value in duration and successive noise samples are uncorrelated.

There are a number of methods by which noise reduction can be achieved. One simple method of reducing low-level noise signals is to set a threshold for incoming samples at some selectable level, such that all values near the minimum detectable signal (MDS) level of the radar are set to zero, while higher-level samples are passed unaltered.

However, noise reduction traditionally has involved averaging, either temporally or spatially. One temporal method averages a number of images to reduce the uncorrelated noise (random signals). This method, known as scan-to-scan integration, was discussed in Chapter 11, and was noted to be a viable method of noise or clutter reduction. A spatial method of noise reduction involves low-pass filtering of the image, thus reducing high-frequency, spiky noise. If the low-pass filtering is implemented by spatial averaging it has the disadvantage of smoothing and blurring the image, particularly around the edges of features. As edges represent a major source of information content for the remote sensing of ice, this technique is undesirable.

A spatial filtering method that reduces this smearing effect is median filtering. In this technique the image is sampled with a moving window. At each pixel the values within the window are arranged in order and the median value is selected and substituted for the original pixel value. Yang and Huang (1985) found that median filtering preserved step- (sharp) edge locations better than averaging, and for impulse noise improved the accuracy of edge location estimation. A second paper by Blom and Daily (1982) found that "median value filtering of radar images removes speckle with minimum edge effects and resolution degradation."

Another smoothing technique, Edge Preserving Smoothing, developed by Nagao and Matsuyama (1979) removes noise in flat regions and does not blur sharp edges. The technique looks for the most homogeneous neighbourhood around each point and then assigns to the point the average grey-level value of the selected neighbourhood. By repeated application of this technique, the noise is reduced while the edges remain sharp. In actual operation a bar mask is rotated around each point. The algorithm detects which mask has the smallest variance of grey-scale levels within the mask and assigns that mask's mean value to the point. This process is repeated until "almost no points in the picture change." To reduce noise, other methods usually take an average over a relatively large region, resulting in the smoothing of small regions. This technique, by iterating many times, is able to achieve noise reduction and only look at small regions (3 x 3 or 5 x 5 pixels). It not only smooths without blurring but, in fact, also sharpens blurred edges. The technique, however, being iterative in nature, is not well suited to real-time display applications.

Histogram Modification

Another group of methods that can provide improvements in the visual quality of an image is histogram modification techniques. A histogram of the pixel values of an image can provide valuable descriptive information regarding enhancement methods that may be applied effectively.

A typical image histogram is shown in Figure 12.1 where 256 grey levels are shown along the X-axis, while the Y-axis shows the frequency of occurrence of each intensity level. It can be seen from this histogram that few of the available grey-scale values are being used. The image from which this histogram was generated would suffer from low contrast.

The problem of limited grey-scale use (limited contrast) can be resolved to a certain extent by a technique known as contrast stretching. With this technique the grey-scale values from the original image are mapped into new values (often by a look-up table) which use the full range of available grey-scale values more effectively.

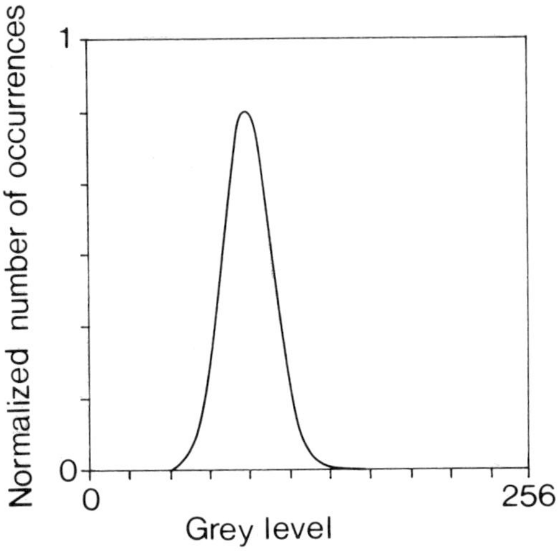

Fig. 12.1 Image histogram showing limited usage of available grey levels. The corresponding image would exhibit limited contrast.

Using this technique it is common to ignore a small percentage of the pixel values by driving them to saturation in the black or white end of the grey-scale to allow for maximum stretching. The basic shape of the histogram, however, is unchanged.

A more effective method of histogram modification is a technique known as histogram equalization. This method improves the visual contrast of an image by adjusting grey-scale values so that each is represented in relatively equal numbers in the image. That is, the histogram of the resulting image is made quite flat. This technique is based on the assumption of a linear human brightness response. This process can be accomplished by calculating the cumulative distribution function (the integral of the original histogram), and then mapping the grey-scale values of the cumulative distribution function onto the desired cumulative distribution function. In the case of simple histogram equalization, the desired distribution function is a linearly-increasing function. The mapped values are then put into a look-up table and by using the re-mapped grey-scale values, the image is equalized.

A further method of histogram modification is a technique called histogram hyperbolization described by Frei (1977) which was based on a similar concept to histogram equalization, except that the histogram was hyperbolized. The basis of this approach is that a model of human brightness perception suggests that the distribution of displayed brightness levels should be hyperbolic. The processing technique should aim for an image with a uniform distribution of perceived brightness levels, rather than an actual uniform distribution of grey levels.

Edge Detection and Enhancement

Another powerful method of image improvement is edge detection or edge enhancement. For the purposes of this study, it has been assumed that edge detection will result in the display of only the edges, while edge enhancement will include some method of adding the detected edge information back onto the original image.

As edges represent abrupt changes in grey-scale level, they have a significant high-frequency content. As such, edge information can be extracted through the use of high-frequency filtering. As noted by Ballard *et al.* (1982) edges are "extremely important; often an object can be recognized from only a crude outline."

The high-frequency filtering can be accomplished in the frequency domain through the use of Fourier transforms. However, this method requires that the whole image be present before processing, whereas other techniques allow operations to take place as the data are being loaded into memory. For the real-time application of this program it is preferable to operate on the image data as they were being received. For example, the high-pass filtering can be accomplished in the spatial domain where a small, square, edge operator is convolved with the image. This technique requires that only a small portion of the image be stored at any one time. Some of these edge operators perform a simple differential function at each pixel of the image, while others provide a measure of the gradient at each point. Because edge operator response varies with edge orientation, operators which are horizontally and vertically sensitive are combined to ensure that all edge information is extracted.

Edge detection involves the application of an edge operator, a comparison of the operator output to a threshold, and a resulting image which contains only edge information. Edge enhancement combines this extracted edge information with the original image to produce a sharpened result.

There are two common classes of edge detectors: local and regional. According to Shaw (1979) "regional operators can be applied on windows several times the width of the edge and can be expected to give responses when the entire region is well described by two constant levels." And again, according to Shaw "regional operators need to find only one edge in a region .... that one may expect such operators to fail if clutter or substantial textural variation is expected."

Since the images to be processed in the ice application are made up of regions with substantial variation and clutter, regional operators were not expected to give useful results and thus were not investigated further.

Local edge detection operators are usually applied on small windows (of the order of 3 x 3 or 5 x 5 pixels), although they can be larger. By applying two directional masks rotated by 90°, the response and an angle of maximum response for the directional local operators can be calculated.

For edge detection the edge operators, which consist of masks of varying weights, are convolved with the image. The masks normally are 2 x 2, 3 x 3, or 5 x 5 pixels, although there are several larger masks referenced in the literature. The output image can consist of the actual operator output values which are related to edge strength, or the operator output can be compared to a threshold and a binary edge/no edge image produced. The threshold can be selected interactively or set to pass a fixed percentage of the strongest edges. Pratt (1978) described an edge-crispening technique, known as discrete convolution filtering, where a high-pass mask was convolved with the image. A method of edge sharpening, without regard to direction, employing a Laplacian mask, was shown in Pratt (1978). This method also convolved the mask with the image.

One of the most often-cited gradient operators (Peli and Malah, 1982) is the Roberts gradient operator, for which the approximation to the gradient is given as

$$G[f(x,y)] \simeq |f(x,y) - f(x+1,y+1)| + |f(x+1,y) - f(x,y+1)| \quad (12.1)$$

and where the pixel orientation is as shown:

f(x,y)	f(x,y+1)
f(x+1,y)	f(x+1,y+1)

The resulting 2 x 2 pixel Roberts operator is:

$$\begin{bmatrix} 1 & -1 \\ 1 & -1 \end{bmatrix}$$

The application of this equation for the enhancement of icebergs from SLAR images is described by Kirby (1981).

Another local operator referenced in the literature (Shaw, 1979) is the Sobel operator. This operator also is an approximation to the gradient, the horizontal mask for which is:

$$\begin{bmatrix} 1 & 2 & 1 \\ 0 & 0 & 0 \\ -1 & -2 & -1 \end{bmatrix}$$

The magnitude of the Sobel operator is evaluated by taking the square root of the sum of the squares of the horizontal and vertical responses.

Another simple operator is the Mero-Vassey operator which in 3 x 3 form is:

$$\begin{bmatrix} 1 & 1 & 1 \\ 0 & 0 & 0 \\ -1 & -1 & -1 \end{bmatrix}$$

The response of this operator is measured as the sum of the absolute values of the vertical and horizontal responses.

There are a number of other operators suggested in the literature and there are several papers which compare these operators. However, the operators seem to perform differently with different types of images such that no operator appears to be universally superior.

Grey-Scale or Colour Table Manipulations

There are several techniques, involving some manipulation of the grey-scale or colour assignment table, that can be used to enhance information in the image. A technique known as "bit slicing" can be used to highlight certain grey-scale levels within the image. By manipulating the display's video look-up table, a small range of grey-scale levels can be assigned a unique grey-scale or colour level, so that features with those grey-scale values will be highlighted in the image. It is possible to have the grey level range of the bit slice controlled interactively such that the highlighted window can be moved through the grey-scale table and the resulting highlighting seen on the image.

As an extension to this technique, density slicing gives all values over a certain grey-scale level a single assigned grey-scale (or colour) level and all values below the cut-off level are assigned to a second grey-scale (or colour) value, resulting in a binary image.

Finally, the introduction of pseudo-colour processing can further enhance images. Pseudo-colour is taken to mean the assignment of a colour, based on an intensity level value, where no colour was present before. The maximum usable number of colours is limited by the spatially diverse nature of the radar image. Adjacent pixels may have substantially different intensities, and to be separable by an operator they must have colour assignments that are readily distinguishable. The application of a few well-chosen colours makes it easier to distinguish between adjacent signal levels, while the assignment of darker colours for low signals and brighter colours for higher signals provides a recognizable progression of intensities.

12.4 TECHNIQUE SELECTION

Although all the techniques discussed in the previous sections improve image quality, some are more suitable for continuous real-time processing of radar images with high data-rates. All the chosen techniques must be capable of being implemented in hardware to be able to deliver the required throughput.

Before presenting the chosen processing techniques, it is useful to review the processing time constraints. The data must be digitized at about 10 MHz, which allows less than 100 ns to perform processing operations between digitized samples. However, because the radar pulse repetition frequency (PRF) is typically 1,000 Hz or less, giving a new set of signal returns every 1 ms, and because in this case only data out to 200 μs (or 30 km, the typical maximum radar horizon) after the transmit pulse are of concern, there are about 800 μs available at the end of each sweep to perform additional processing.

It is also important to define real-time in the context of this research. "Real-time" is understood to be the ability to handle the data out to 30-km range on a continuous basis without having to miss incoming data. Image enhancement will be accomplished by taking advantage of the dead time at the end of each sweep.

To provide a basis for comparison for subsequent results, Figure 12.2 shows a photograph of the original radar data on a typical PPI display. To permit fair comparison of the images, the contrast of each of the following images has been adjusted to use the full available grey scale.

Figure 12.3 shows a photograph of the digitized data displayed on a high-resolution display system with 1,280 x 1,024 pixels. (Because of the digitizing hardware available, the data were digitized to 8 bits at 11 MHz.) To produce this image, the scan conversion

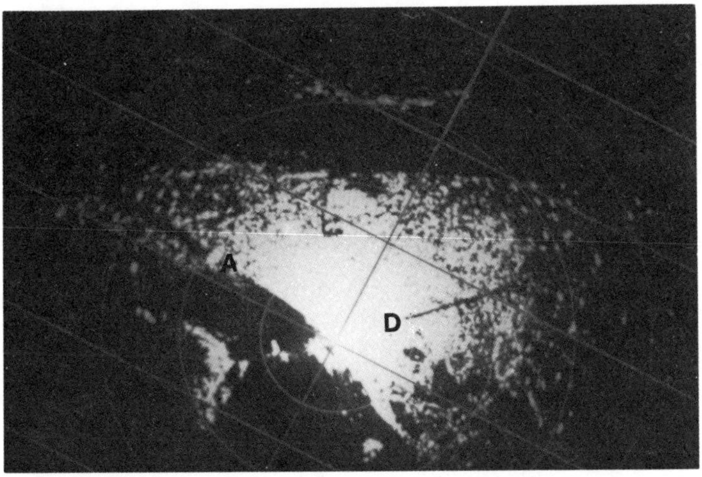

Fig. 12.2 Original radar image on PPI display. Iceberg A shows a resolvable return while target D, Adams Island, is only detectable because of its radar shadow.

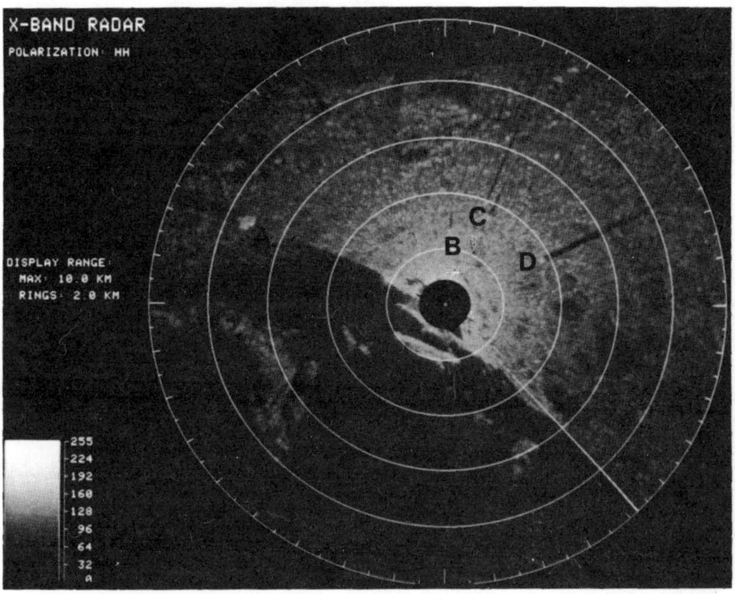

Fig. 12.3 Radar image on digital display. Iceberg targets at A, B, and C are now clearly resolved as is the island at D.

routine retained the last polar value assigned to each pixel on the X-Y display. This technique has the disadvantage that small isolated targets can be lost as they are replaced by a subsequent lower-level signal. For the application of displaying imagery of an ice-covered region in which the targets are spatially large, this would not be a major problem. However, for the detection of singular targets in ice-infested waters a more elaborate technique, possibly using a peak detector, would be required.

Examination of the differences in Figures 12.2 and 12.3 shows the image improvement afforded by the wider dynamic range of the raster scan display. The individual targets such as the icebergs at A, B, and C, are lost in the saturated video of the PPI display, but are clearly shown on the raster scan display. If the PPI display intensity is turned down, small, but significant, targets and targets at far range are lost.

To reduce background noise, the incoming data were subjected to a threshold level. The threshold level was determined by performing a histogram on the system noise in a non-signal area of the image and selecting the level that rejected 95% of the noise samples. This process substantially reduced system noise. Care must be taken in choosing the threshold value so that small targets are not lost. If the thresholding is not performed prior to range normalization, the R^3 normalization weighting causes the noise level to increase with range, and this is distracting to the observer.

The next operation carried out was range normalization, which eliminated the effect of the range factor in the radar equation and displayed targets of equal radar cross-section at equal intensity regardless of range. Figure 12.4 shows the results of applying a $1/R^3$ normalization.

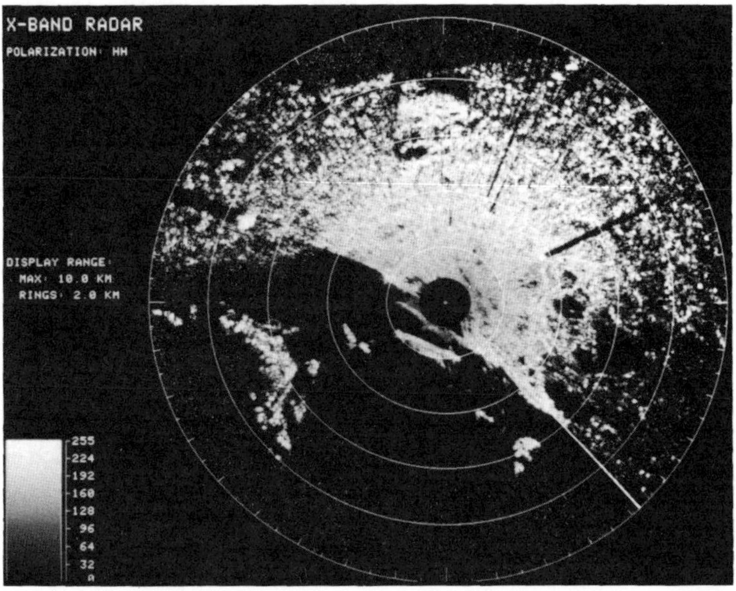

Fig. 12.4 Range normalized image. Image now exhibits uniform brightness independent of range.

In comparing Figure 12.4 with the raw radar image it can be seen that the normalized image has a superior appearance, simply because there is a uniform intensity across the image. An examination of the figure, however, shows a considerable amount of noise, particularly at far range.

The choice of a smoothing (filtering) routine to remove the noise was more difficult. The routine that yielded the best results, Edge Preserving Smoothing, requires multiple passes to be effective and, as such, cannot be implemented in real- or near real-time. The median filter, however, yielded acceptable results and can be implemented relatively easily in hardware to operate in real-time.

As the median filter removes 'spikes' that are shorter than one half the filter length, the filter length must be chosen so that it will not eliminate single targets. At the 11-MHz digitizing rate there are two samples per pulse length and a minimum of two returns from a single point target; therefore, a filter cannot be used which is larger than 3 x 3 pixels. Figure 12.5 shows the results of applying the median filter.

The next process implemented was edge detection and enhancement. A review of edge detection techniques showed that the Mero-Vassey operator performed as well as any tested for this application. As it can be implemented using only binary operations, it is suitable for hardware implementation. Figure 12.6 shows the detected edges after performing the threshold process. However, when the edges were added back to the image shown in Figure 12.5, there was no improvement in image interpretability. Probably this occurred because of the sharp contrast already in the image. An improvement in image quality by adding the edges back in colour can be achieved but unfortunately cannot be shown in this text.

Further substantial improvements in image interpretability are possible through the use of colour. In particular, the use of a grey scale topped by two different colour slices and with the edge information added in a third colour substantially enhances the image interpretability.

12.5 HARDWARE CONSIDERATIONS

This section discusses some necessary considerations when implementing (in hardware) the techniques selected in the previous section.

The analogue radar video is digitized using a readily available single-chip A/D converter. The converter output is passed through a comparator circuit whose

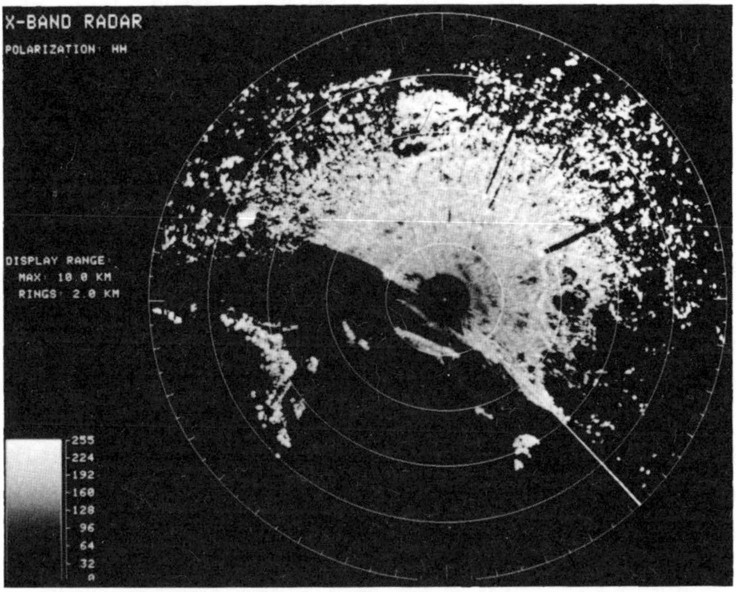

Fig. 12.5 Median filtered image. Note marked reduction in "speckle" noise, particularly at far range.

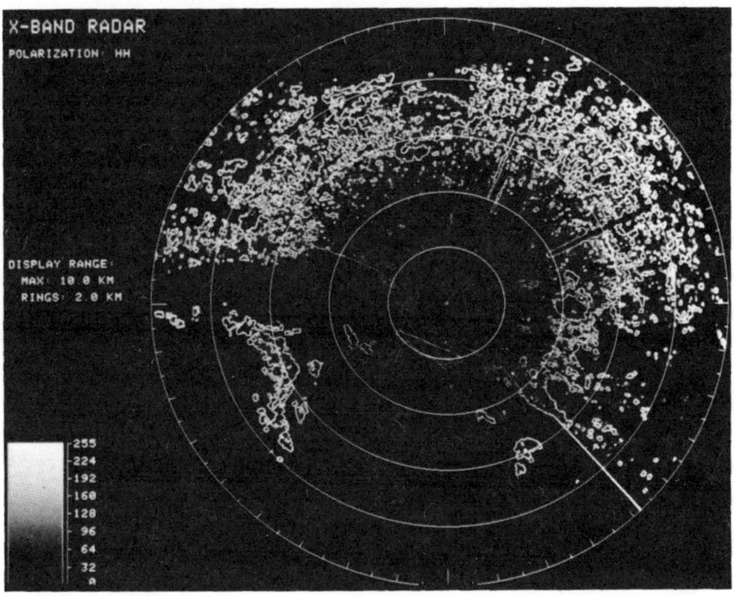

Fig. 12.6 Edge detected image.

output is zero for input values below the operator-controlled threshold. Values above the threshold are passed unchanged. The data are then normalized in range through multiplication by a range-dependent constant taken from a read-only memory (ROM). Each range position has a different multiplier value in ROM.

To minimize the hardware, the 3 x 3 pixel median filter is implemented sequentially in range then bearing, using two three-point, one-dimensional filters. The range filtering is done along each sweep as the data are received. The range filter output is fed into a three-sweep buffer. The buffer is then accessed to provide the output for the three-point bearing median filter.

The bearing filter output is passed on to the edge detection and enhancement circuitry. The use of a 3 x 3 pixel edge operator requires another three-sweep buffer. Since the Mero-Vassey operator coefficients are 0, 1, and -1, multiplication reduces to addition and subtraction. The horizontal and vertical operator magnitudes are combined to produce the edge operator output. For an image of detected edges, this output is subjected to a threshold and passed to the scan converter for display. For edge-enhanced images, the edge operator output is combined in a weighted sum with the buffered median-filter output. This sum is passed on to the scan converter for display. With more complex circuitry providing programmable multipliers, any 3 x 3 convolution operator can be implemented. Alternatively, the edge image can be treated as a colour overlay through manipulation of the colour look-up table.

For scan conversion, the circuitry has the basic purpose of storing the range-bearing sample in the appropriate X-Y display location (pixel). The mapping from range-bearing sample co-ordinates to X-Y location can be implemented through look-up table techniques using ROM.

To reduce storage requirements, the look-up table, because of symmetry, need only contain the mapping for a single half quadrant.

Assuming the use of a 1,024 line by 1,024 pixels per line display, there is a maximum of 512 pixels (range positions) per sweep. To allow one look-up table to suffice for various range settings on the display, the number of incoming data points per sweep fed to the scan converter must be maintained at 512. For the 10-km range coverage at the 10-MHz sampling rate, there are about 1,024 samples per sweep. To reduce this to the required 512 displayable points, with minimum loss of information, successive pairs of samples are averaged. When the maximum range being displayed is reduced to 5 km, the first 512 sample points are used. As the maximum range being displayed is reduced further by factors of two, the appropriate data points are replicated to maintain an input of 512 data points to the scan converter.

Raster scan displays are presently available in a 1,280 pixel by 1,024 line format. These provide a square 1,024 x 1,024 area for the radar image and a side area for annotation. Each screen pixel typically has an 8-bit storage location associated with it. This 8-bit pixel value is used to index a video look-up table whose output determines the colour of the pixel, by controlling the intensity of each of the red, green, and blue guns. Most colours can be produced by an appropriate mixture of red, green, and blue, while for grey scale the red, green, and blue intensities at each level are set equal. The video look-up table is a powerful tool in that the appearance of the image can be changed without altering the original data. For example, such techniques as bit slicing and pseudo-colour mapping can be implemented in this manner.

12.6 SUMMARY

This chapter looked at radar display improvements and image processing techniques which substantially improve the interpretability of ice radar images. The chapter outlined several major deficiencies in existing marine radar displays and indicated how these can be eliminated with a digital display. The chapter then reviewed a number of simple image processing techniques which might be suitable for the display of ice images. A number of these techniques which are suitable for real time implementation were investigated and images demonstrating the results were shown and discussed. Finally, several hardware considerations involved in the implementation of these techniques were discussed.

CHAPTER 13
Ice Radar Systems Design

13.1 REVIEW

The results of the experimental and theoretical discussions contained in the previous chapters, which are pertinent to the design of an ice-detecting radar system, are now summarized. It is these findings that can be used to guide the selection of the design specifications for such a radar system.

In Chapter 3, the physical and electrical properties of multiyear ice and glacier ice were shown to differ from each other and from those of first-year ice. Multiyear ice and glacier ice are known to be a threat or impediment to navigation. Fortunately, the electrical properties of multiyear ice and glacier ice are significantly different from first-year ice, thus offering a potential for ice type differentiation.

Chapter 6 showed that the multipath phenomenon can be a problem both over an ice cover and over open water. Multipath interference results in a frequency-dependent vertical lobing of the antenna pattern such that lower frequency radar systems are unable to direct significant energy onto the surface. However, more energy can be made incident on the surface by increasing the frequency. Attenuation of the radar signal by fog or precipitation at X- and S-bands at the short ranges involved in this program is, in general, not significant. However, changes in propagation conditions can adversely affect target detection and these degradations were shown to be particularly apparent for subrefractive conditions that can occur with high frequency off the east coast of Canada. Chapter 6 concluded with a look at sea clutter and its effect on target detection. It was noted that high-frequency sea clutter decorrelates quickly and is not anticipated to be a significant problem in target detection. However, the longer-duration sea spikes demonstrate target-like characteristics for relatively long periods of time. Finally, it was shown that for low and moderate sea clutter, horizontally-polarized returns were much lower than vertically-polarized returns, while at higher sea states the horizontal and vertical returns are expected to be similar.

In Chapter 7, it was shown that ice returns at X- and K_u-band were usually cluttered, with energy being returned from targets that were not of navigational interest. The returns from S-band, on the other hand, were correlated with features having substantial physical extent. The K_a-band returns indicated an improved capacity to distinguish multiyear ice and, to a lesser extent, glacier ice.

Chapter 8 presented some of the most substantial results of the research program. It was shown that for a full ice cover, target-to-clutter improvements of 10 dB could be achieved for multiyear ice by employing the cross-polarized radar returns. For icebergs, the improvements in target-to-clutter performance varied from a gain of as much as 17 dB to a loss of 3 dB. Returns from even substantial first-year pressure ridges were almost eliminated on the cross-polarized channel. The cross-polarized returns from both horizontal and vertical transmit were essentially the same. For the open water case, there was no apparent improvement in target-to-clutter ratio provided by the cross-polarized return, as both the target (iceberg) and the clutter (sea) essentially depolarized the radar signal to a similar extent. It was shown, however, that because of the nature of the variability of the like- and cross-polarized returns, substantial improvement in the probability of detection could be achieved through the use of both radar channels.

As sea clutter returns were minimal with horizontal transmit, and the cross-polarized returns from the ice target for horizontal or vertical transmit were equal, the choice of horizontal transmit as the preferred polarization is confirmed.

In Chapter 10, the choice of the highest possible placement of the radar antenna was confirmed based on maximum range of detection and overcoming multipath interference, although a dual antenna system at a high and low elevation may be optimum under some conditions (ducting). It was also noted that although increased azimuthal and range resolution reduced the mean clutter power, thus improving the target-to-clutter ratio, the resulting increase in the tails of the clutter distribution introduced additional difficulties into the detection process.

Chapter 11 showed that improvements in target detection can be achieved by employing signal processing techniques such as integration. The integration can be done on both a pulse-to-pulse and scan-to-scan basis. The advantages of a constant false alarm rate processor were also discussed.

Finally, Chapter 12 showed improvements that can be achieved in interpreting ice images by employing a wide dynamic range raster-scan display system and several simple image processing algorithms.

13.2 SYSTEM CONFIGURATION

The optimum system configuration for the ice-classification application and the open-water, ice-detection application consist of the same basic radar hardware. The two applications would differ, however, in their signal processing requirements. A suggested design of a radar system for ice classification will be presented first, followed by a description of the additional requirements for an open-water ice detection system.

Ice Classification

The ice classification system should consist of two radar systems: (i) an S-band system for the detection of large targets and (ii) a dual-polarized, X-band system for classification of icebergs and multiyear ice. To provide coincident radar returns for an integrated display, the X- and S-band antennas should be collocated and rotate as one. Dual X/S-band antenna units are commercially available but require modification to accommodate the required dual-channel, X-band rotary joint and the vertically-polarized X-band antenna. (A version of such a vertically polarized antenna was developed by the Canadian Department of Fisheries and Oceans under contract, and is about 3.8 m long by 25 cm high and 60 cm deep.)

To avoid possible interference between the X- and S-band radars, the radars should transmit alternately. The X-band radar should be configured to transmit horizontal polarization and simultaneously receive both horizontal and vertical polarizations. The three channels of radar video should be simultaneously digitized and scan converted. Each channel can then be image processed, consisting of range normalization, noise reduction, and possibly some form of edge enhancement. The three channels of radar data can then be integrated in an operator-selected fashion on a single raster scan display device.

Research into the most effective method of achieving the display integration on a single display is still underway. One interim solution is to use one display for the S-band image, and a second display for the X-band like-polarized image on which the cross-polarized image is overlaid using colour.

Open-Water Detection

The open-water detection system should employ the same dual-polarized, X-band radar and antenna system as discussed in the previous section. Both channels should independently undergo constant false alarm rate (CFAR) processing and then pulse-to-pulse integration. The data can be processed in a ground-stabilized mode to remove the effects of ships' motion. The two channels then can be scan converted independently, scan-to-scan integrated, and fed to one raster scan display. They can be combined such that a target will be displayed if it is detected in either channel. The detection threshold can be set by the operator or be determined automatically by the display.

13.3 SUMMARY

This chapter began with a summary of the experimental results as they apply to the design of specialized ice radar hardware. The chapter concluded with descriptions of ice radar systems configurations for the ice classification environment and for the open-water detection of icebergs.

CHAPTER 14
Ice Radar System Demonstration

14.1 EQUIPMENT INSTALLATION

During late 1985 and early 1986 the elements of a dual-polarized X-band radar system were integrated into an operational unit. In April of 1986, the radar system was installed on the Canadian ice strengthened carrier MV *Arctic*.

The radar system comprised two conventional 25-kw radar systems and a custom turning unit and antenna system. The antenna system consisted of separate collocated horizontally- and vertically-polarized antennas. One of the radar systems was connected to the horizontally-polarized antenna and was configured to transmit and receive horizontally-polarized energy. The second radar, configured for receive-only, was connected to the vertically-polarized antenna and displayed vertically- (cross-) polarized returns. The radars fed identical conventional PPI displays (Figure 14.1) which were mounted side-by-side on the forward bridge of the ship. This arrangement allowed comparison of the like- and cross-polarized signals and easy access for the mariner.

The antenna system, Figure 14.2, consisted of a conventional 9' X-band, horizontally-polarized marine antenna (top of the figure) and a specially designed and

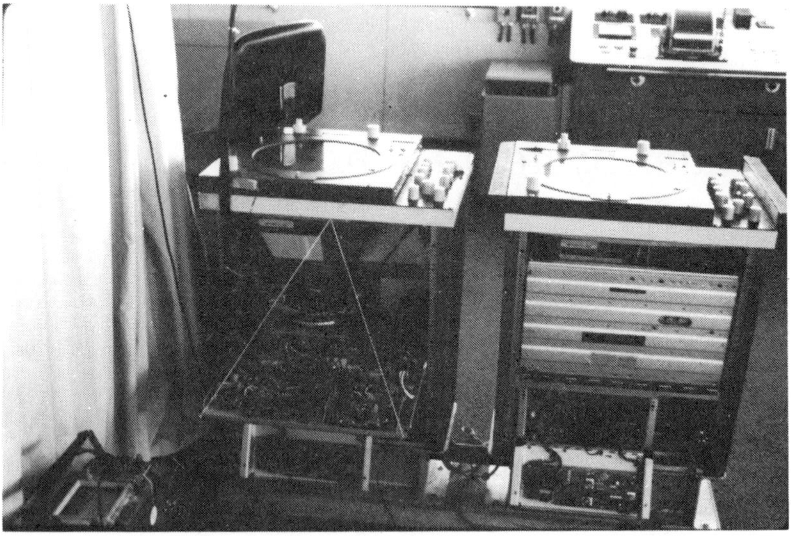

Fig. 14.1 Like- and cross-polarized PPI displays.

Fig. 14.2 Dual-polarized X-band antenna.

constructed vertically-polarized antenna (bottom of the figure). The vertically-polarized antenna was designed and constructed by the Department of Electrical Engineering at the University of Manitoba. The horizontally-polarized antenna had a 0.8° azimuthal beamwidth and a 20° vertical beamwidth, while the vertically-polarized antenna had a 1.0° azimuthal beamwidth and a 15° vertical beamwidth. The two antennas were rotated as one by a modified S-band turning unit which contained a dual-channel rotary joint.

During the course of research voyages carried out in June and November-December of 1986, the radar equipment described above was augmented with data acquisition equipment and a radar digitizer/scan converter and digital display (Figure 14.3). The digitizer is on the right side of the figure (at A), while the scan converter is located in the centre of the figure (at B). The scan-converted digital images were displayed on a monochrome and a colour monitor, shown on the left of the figure.

Fig. 14.3 Data acquisition and digital display system.

The radar data were acquired and stored on an IBM-AT computer.

14.2 THE TEST PROGRAM

The equipment was first operated during a June 1986 crossing of the MV Arctic from Antwerp, Belgium, to Nanisivik on Baffin Island. However, a number of basic radar problems prevented the successful testing of the equipment during this trip. The most serious problem was water in the waveguide, particularly in the dual-channel rotary joint. The water entered the waveguide and rotary joint due to improper installation of the waveguide by the contractor during initial installation. Following this trip, the radar system, including the full antenna system, was removed from the ship and completely overhauled. Prior to the next trip, the radars and antenna system were extensively tested, and the waveguide and rotary joint in the turning unit were replaced. The radar system was reinstalled on the ship in mid-November in Antwerp, including the installation of completely new waveguide runs.

The testing of the radar system was completed during late November and early December of 1986. The MV Arctic departed Antwerp bound for Nanisivik on Baffin Island and late-season ice trials. The ship encountered a complete cover of first-year ice north of Disko Island on the west coast of Greenland. Mixed in with the first-year ice were a large number of icebergs, bergy bits and growlers. Near the entrance to Lancaster Sound, the ship encountered a high concentration of multiyear ice contained in a full cover of first-year ice. In fact, the multiyear ice was so heavy that the ship was unable to make its destination of Nanisivik and had to turn back without completing some of its test program.

Data Collection

During the course of the voyage a large quantity of dual-polarized radar data were collected from sea clutter, icebergs in open water, icebergs in a full ice cover, and multiyear ice in a full ice cover. The radar data were collected in two forms. First, data were taken for selected targets within a sample window that was either 64 samples in range by 128 samples in azimuth, or 16 samples in range by 512 samples in azimuth. With a digitizing rate of 5 MHz, the samples in range were 30 metres apart. The data collection system simultaneously acquired like- and cross-polarized returns digitized to 8-bit resolution. The system collected samples from up to thirty consecutive scans. This part of the data acquisition system was identical to that used in the 1984 experiment described earlier. The second data acquisition method consisted of freezing the display frame of the scan converter and then transferring the image data on a GPIB bus to the computer. Because of the single channel nature of the scan converter, like- and cross-polarized images had to be sequentially captured.

During the early portion of the experiment, data were collected from several targets that were visually sighted, described, and photographed. However, during a major part of the experiment the ship was operating in 24-hour darkness, and only targets that were close to the ship and within range of the ship's search lights could be documented.

14.3 OPERATIONAL USE

After a short introductory period, the radar system was extensively used by the officers of the ship to navigate in difficult ice situations. While navigating through a high concentration of icebergs and bergy bits in a full ice cover, the cross-polarized display provided a vastly superior indication of icebergs and bergy bits. This improvement is shown in Figure 14.4, where the image at the top is the output of the digital display system of the like-polarized channel, and the image at the bottom is the cross-polarized image from the same display. The range from the centre to the edge of the image is 3 nm. Note the substantial improvement in display of targets at points A through O. The smallest target detected with the cross-polarized display during the voyage was a 1.5 m by 15 m growler, which was initially displayed at a range of 2 nm. The target was not distinguishable on the like-polarized display and was only visually detected at about 3/4 nm.

As the cross-polarized image of Figure 14.4 shows, there is still a substantial amount of return energy out to about 1/2 nm from the ship, and which is not necessarily associated with hazardous targets. These high returns result from a different scattering mechanism, due to the steeper grazing angles in this range and to saturation of the display at close ranges because of the strong range dependency of the returned power. The operational procedure used during the voyage was to set the gain of the cross-polarized display to give clutter return out to about 1/2 nm and then to declare hazardous any target that was detected at a range greater than 1 nm. This technique resulted in no false target declarations and no known target misses.

The second scenario under which the radar provided invaluable information was in navigating in a field of multiyear ice frozen into a full first-year ice cover. In this situation, the radar system performance was even more impressive. The cross-polarized display was able to clearly display and delineate multiyear ice floes. This is shown in Figure 14.5, where the image at the top is from the like-polarized radar, while the image at the

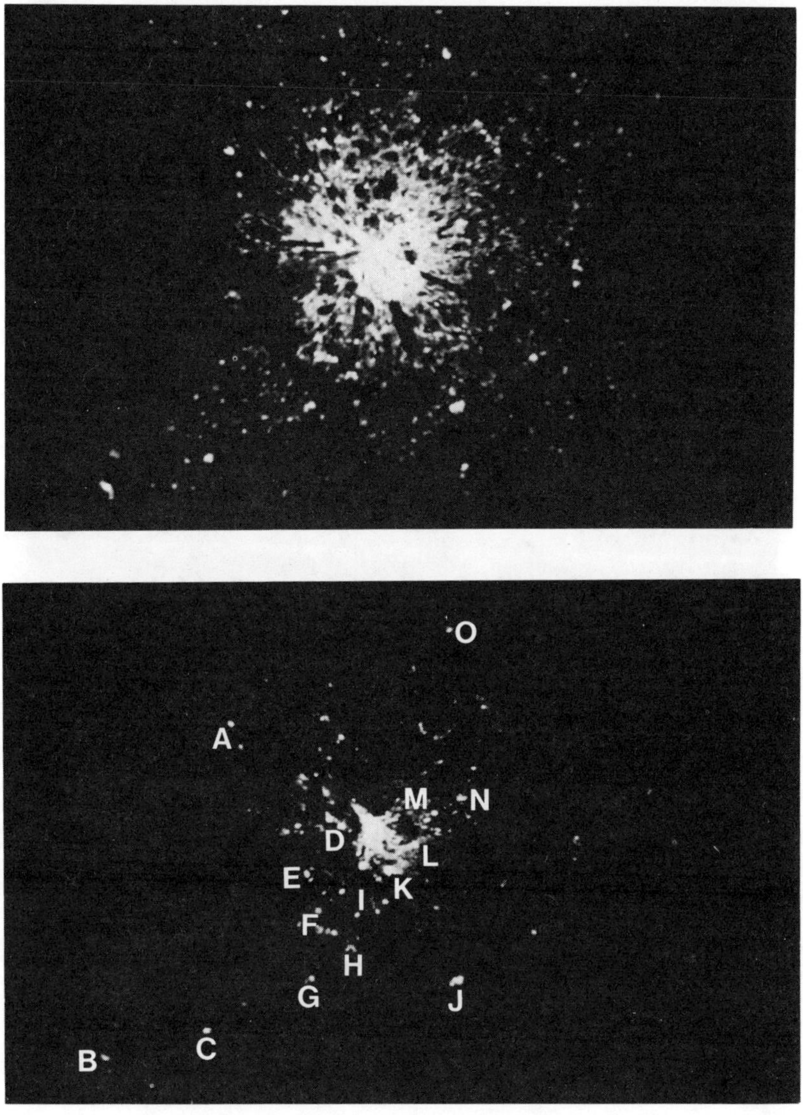

Fig. 14.4 Like-(top) and cross-polarized (bottom) images of icebergs in a full first-year ice cover. Icebergs at A through O.

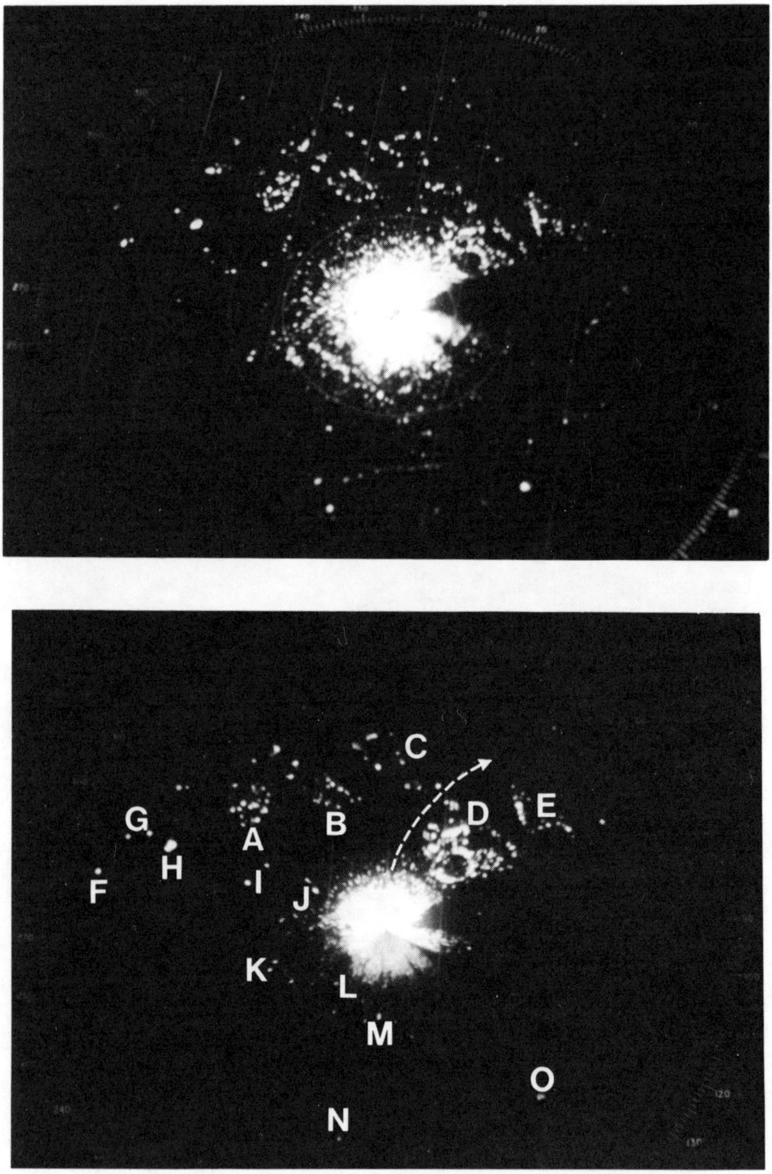

Fig. 14.5 Like-(top) and cross-polarized (bottom) images of multiyear ice floes and icebergs in a full first-year ice cover. Multiyear ice at A through E and icebergs at F through O.

bottom is from the cross-polarized radar. These images are photographs of the PPIs of the two radars. The range from the centre to the edge of the images is 6 nm. In addition to clearly showing the multiyear floes at points A through E, the cross-polarized display also showed the icebergs at F through O. Using the cross-polarized display, the navigator was able to maneuver the ship up through the gap in the multiyear ice as shown in the cross-polarized image. This route would not have been determinable with the conventional image at the top. The operational procedure for the radar in this environment of heavier first-year ice was to set the cross-polarized radar gain to display clutter out to about 1 nm, and to declare any target displayed at a range of greater than 1.5 nm to be hazardous. Again, no targets were falsely declared as hazardous, nor were any known hazardous targets missed.

Once the cross-polarized radar settings were established for the two ice regimes, the settings were left constant for operational use. It should be emphasized that no matter how the settings of the like-polarized display were manipulated, performance anywhere approaching that of the cross-polarized display could not be achieved. Further operational use, however, may suggest gain settings for the cross-polarized display which will yield improved operational performance.

The radar system is still installed on the ship where it will remain for the 1987 navigational season, in order to provide extensive operational experience.

14.4 DISCUSSION

The use of the dual-polarized radar, and in particular, the cross-polarized display provided substantial improvements in the ability to navigate in ice fields containing icebergs and multiyear ice.

However, during the experimental program it was discovered that there was still a slight pattern misalignment between the horizontally-polarized and the vertically-polarized antennas. This resulted in reduced returns in the cross-polarized channel and reduced performance of the cross-polarized display. In addition, the gain of the cross-polarized antenna is below its specified value due to damage to the antenna caused by transmitting into it during the 1986 spring experiment. This again will have resulted in degraded performance of the cross-polarized display. The antennas are being removed and repaired during the winter season and will be reinstalled prior to the 1987 arctic shipping season. These repairs and realignment are anticipated to provide a 5 to 8 dB improvement in the sensitivity of the cross-polarized radar.

During the course of the experiment and subsequent analysis, the performance of the digital display system did not live up to expectations. Where it had been anticipated that the digital display would provide improved performance over the PPI display, this did not prove to be the case. In fact, as shown by Figure 14.6, where the top image is a photograph of the PPI display, and the bottom image is a photograph of the digital display for the same ice scene, it can be seen that the PPI provided a clearer definition of the multiyear floes. However, this was the first occasion for real-time use of the display. Efforts on this voyage concentrated on raw data collection. Subsequent efforts will address improving the digital display.

14.5 FUTURE WORK

Although the ice radar problem has received considerable attention, several areas requiring additional work remain. In particular, much remains to be done to optimize

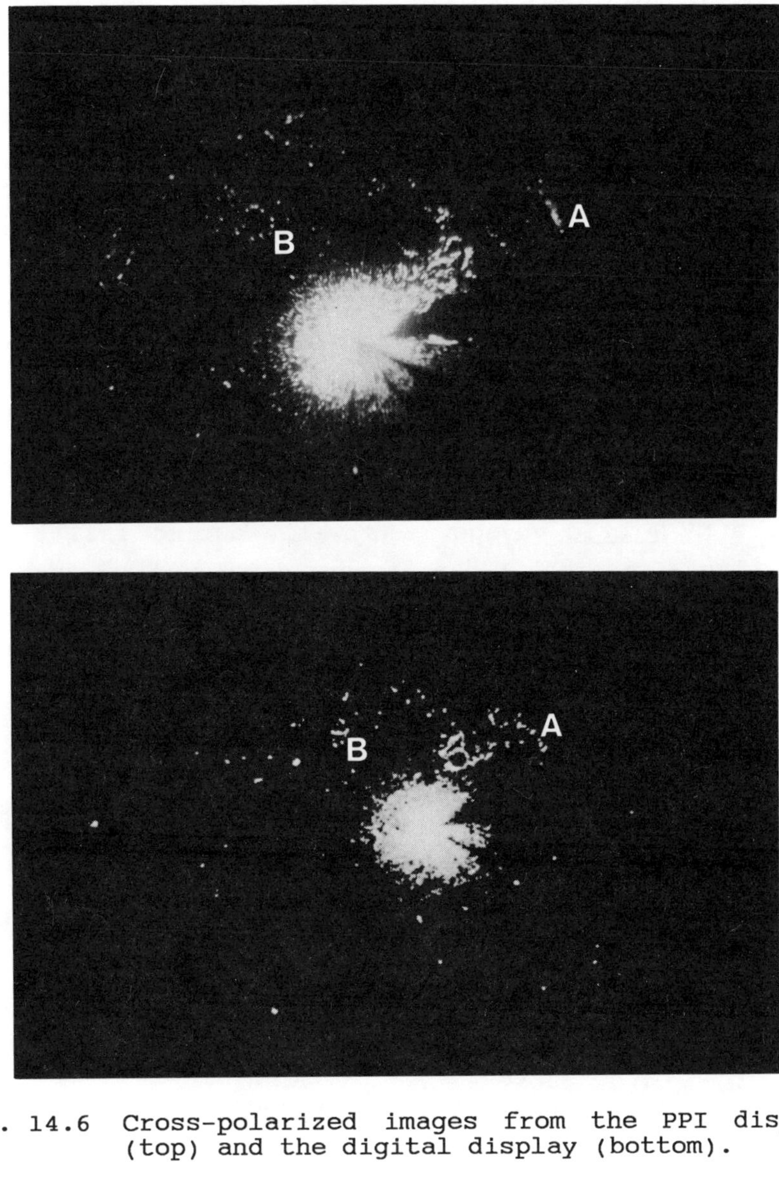

Fig. 14.6 Cross-polarized images from the PPI display (top) and the digital display (bottom).

the display of the radar data. Some additional experimental work with higher frequency systems and on target detection in open water is required.

Work is continuing on devising the most effective method of displaying the radar data. The system installed on the MV *Arctic* is using a single channel, commercial scan conversion system. The like- and cross-polarized channels require manual switching. The existing display system is also monochrome only. An extension to this display system is currently being designed which will allow for the mixing of the radar channels by addition, subtraction or ratio of the inputs. In addition, the system will be in colour and will allow one radar channel to be overlaid on the other in a different colour.

More data on the detection of ice targets in open water are required in order to predict detection performance. Some of this data will hopefully be collected during a future MV *Arctic* voyage. As well, during 1987 a dual-polarized, coherent X-band radar, developed and constructed at McMaster University, will be tested at a coastal site in Newfoundland. Data will be collected for sea clutter and for icebergs.

There is also a need to collect more data on the return from ice at K_a-band, in order to quantify the possible improvements that may be obtained at higher frequency.

The experimental work reported in this book, culminating in the successful demonstration of a dual-polarized ice radar, indicates that significant improvements in the use of surface-based radar for ice surveillance are indeed possible. Further work is now needed to realize even greater gains in ice-radar performance. It is hoped that this book has provided a firm research foundation on which these future gains can be based.

Despite the rigours of arctic field programs, northern researchers are always impressed by the breathtaking grandeur of the North.

(View of Bylot Island from the Radar Camp)

Bibliography

Addison, J.R. 1969. Electrical properties of saline ice. Journal of Applied Physics, Vol. 40, No. 8, 3105-3114.

Almond, T. and J. Clarke. 1983. Consideration of the usefulness of microwave propagation prediction methods on air-to-ground paths. Proceedings of the Institution of Electrical Engineers, Vol. 130, Part F, No. 7, 649-656.

Anderson, D.L. 1960. The physical constants of sea ice. California Institute of Technology, Research, Vol. 13, No. 8, 310-318.

Anderson, K.D. 1983. Surface-search radar performance in the evaporation duct: global predictions. Naval Ocean Systems Center, Technical Report 923, San Diego, California.

Anderson, V.H. 1960. High altitude, side-looking radar images of sea ice in the Arctic. 4th International Symposium on Remote Sensing, Ann Arbor, Michigan, 845-857.

Andrews, H.C. 1974. Digital image restoration: survey. Computer, Vol. 7, No. 5, 36-45.

Arakawa, K. and K. Higuchi. 1954. On the freezing process of aqueous solution. Low Temperature Science, A12, 73-86 (in Japanese).

Assur, A. 1958. Composition of sea ice and its tensile strength. Arctic Sea Ice; U.S. National Academy of Sciences-National Research Council, Pub. 598, 106-138.

Bailey, E.B. and N.C. Randall. 1961. Detecting signals in noise. Electronics, Vol. 34, No. 11, 66.

Ballard, D.H. and C.M. Brown. 1982. Computer vision. Prentice Hall, Englewood Cliffs, N.J.

Baumgartner, G.B. Jr., H.V. Hitney, and R.A. Pappert. 1983. Duct propagation modelling for the integrated-refractive-effects prediction system (IREPS). Proceedings of the Institution of Electrical Engineers, Vol. 130, Part F, No. 7, 630-642.

Beckmann, P. and A. Spizzichino. 1963. The scattering of electromagnetic waves from rough surfaces. Pergamon Press, Oxford.

Bell, J. 1967. Propagation measurements at 3.6 and 11 Gc/s over a line-of-sight radio path. Proceedings of the Institution of Electrical Engineers Vol. 114, No. 5, 545-549.

Benedict, P. and J. Hall. 1979. A study of iceberg detection for marine navigation. Transport Canada, Ottawa, Report TP 2409.

Blake, L.V. 1969. A guide to basic pulse-radar maximum-range calculation Part 1 - equations, definitions and aids to calculation. Naval Research Laboratory, Washington, D.C., Report 6930.

Blake, L.V. 1980. Radar range-performance analysis. Lexington Books, Lexington, Massachusetts.

Blanchard, A.J. and J.W. Rouse, Jr. 1980. Depolarization of electromagnetic waves scattered from an inhomogeneous half space bounded by a rough surface. Radio Science, Vol. 15, No. 4, 773-779.

Blom, R.G. and M. Daily. 1982. Radar image processing for rock type discrimination. Institute of Electrical and Electronic Engineers Transactions on Geoscience and Remote Sensing, Vol. GE-20, No. 3, 343-351.

Budinger, T.F., R.P. Dinsmore, P.A. Morill, and F.M. Soule. 1960. Iceberg detection by radar. International Ice Patrol Bulletin No. 45, United States Coast Guard, 49-97.

Campbell, W.J., J. Wayenberg, J. Ramseyer, and R.O. Ramseier. 1978. Microwave remote sensing of sea ice in the Aidjex main experiment. Boundary-Layer Meteorology, Vol. 13, 309-337.

Cloude, S.R. 1983. Polarimetric techniques in radar signal processing. Microwave Journal, July, 119-127.

Cox, G.F.N., J.A. Richter-Menge, W.F. Weeks, and M. Mellor. 1984. Mechanical properties of multi-year sea ice; Phase 1: test results. U.S. Army Corps of Engineers, Cold Regions Research and Engineering Laboratory, Hanover, N.H., Report 84-9.

Cox, G.F.N. and W.F. Weeks. 1974. Salinity variations in sea ice. Journal of Glaciology, Vol. 13, No. 67, 109-120.

Croney, J. 1966. Improved radar visibility of small targets in sea clutter. The Radio and Electronic Engineer, September, 135-147.

Croney, J., A. Woroncow, and B.R. Gladman. 1975. Further observations on the detection of small targets in sea clutter. The Radio and Electronic Engineer, Vol. 45, No. 3, 105-115.

Cumming, W.A. 1952. The dielectric properties of ice and snow at 3.2 centimetres. Journal of Applied Physics, Vol. 23, No. 7, 768-773.

Currie, B.W. and E.O. Lewis. 1985. Evaluation of dual-polarization in the surface-based radar detection of icebergs. Proceedings of the 8th International Conference on Ports and Oceans Engineering Under Arctic Conditions, Narssarssuaq, Greenland, 757-766.

Davidson, L.W. and W.W. Denner. 1982. Sea ice and iceberg conditions on the Grand Banks affecting hydrocarbon production and transportation. Proceedings of Institute of Electrical and Electronic Engineers, Oceans '82, 1236-1241.

Dougherty, H.T. and B.A. Hart. 1979. Recent progress in duct propagation predictions. Institute of Electrical and Electronic Engineers Transactions on Antennas and Propagation, Vol. AP-27, No. 4, 542-548.

Ebbesmeyer, C.C., A. Okubo, and H.J.M. Helset. 1980. Description of iceberg probability between Baffin Bay and the Grand Banks using a stochastic model. Deep-Sea Research, Vol. 27A, 975-986.

Ellis, R.N. 1984. Theoretical study of radar scattering from an iceberg. MSc Thesis, McMaster University, Hamilton, Ontario.

Encyclopaedia Britannica. 1985. 15th Edition, Vol. 20, 813.

Evans, S. 1965. Dielectric properties of ice and snow - a review. Journal of Glaciology, Vol. 5, No. 42, 773-791.

Ewell, G.W., M.T. Tuley, and W.F. Horne. 1984. Temporal and spatial behaviour of high resolution sea clutter spikes. Proceedings of the 1984 Institute of Electrical and Electronic Engineers, National Radar Conference, Atlanta, Georgia.

Fay, F.A., J. Clarke, and R.S. Peters. 1977. Weibull distribution applied to sea clutter. Institution of Electrical Engineers International Radar Conference, London, England, 101-104.

Frankenstein, G. and R. Garner. 1967. Short note: equations for determining the brine volume of sea ice from -0.5 to -22.9° C. Journal of Glaciology, Vol. 6, No. 48, 943-944.

Frei, W. 1977. Image processing by histogram hyperbolization. Computer Graphics and Image Processing, Vol. 6, 286-294.

Fung, A.K. 1967. Theory of cross polarized power returned from a random surface. Applied Science Research, 18, 50-60.

Fung, A.K. and H.J. Eom. 1982. Application of a combined rough surface and volume scattering theory to sea ice and snow backscatter. Institute of Electrical and Electronic Engineers Transactions on Geoscience and Remote Sensing, Vol. GE-20, No. 4, 528-536.

Gammon, P.H., R.E. Gagnon, W. Bobby, and W.E. Russell. 1983. Physical and mechanical properties of icebergs. 15th Annual Offshore Technology Conference 4459, Houston, Texas, 143-149.

Glaser, H.C. and T.F. Havig. 1981. Radar backscattering from ocean waves at low grazing angles. Institute of Electrical and Electronic Engineers National Aerospace and Electronics Conference, Vol. 1, 406-411.

Gogineni, S., R.G. Onstott, R.K. Moore, and J. Chancellor. 1984. Intermediate results of radar backscatter measurements from summer sea ice. Remote Sensing Laboratory, University of Kansas, Center for Research, Inc., Lawrence, Kansas, Technical Report 3311-2.

Gow, A.J. 1968. Bubbles and bubble pressures in Antarctic glacier ice. U.S. Army Corps of Engineers, Cold Regions Research and Engineering Laboratory, Hanover, N.H., Research Report 249.

Hagman, T., J. Nilsson, and Y. Nilsson. 1976. Sea ice-75: Flar, Odar, ship's radar. Swedish/Finnish Winter Navigation Research Board Report No. 16:5.

Hansen, G.V. 1973. Constant false alarm rate processing in search radars. Institution of Electrical Engineers International Radar Conference, London, England, 325-332.

Haykin, S. and B.W. Currie. 1984. A phased array radar for iceberg detection. Communications Research Laboratory, McMaster University, Hamilton, Ontario, CRL Report No. 121.

Haykin, S., B.W. Currie, E.O. Lewis, and K. Nickerson. 1985. Surface-based radar imaging of sea ice. Proceedings of the Institute of Electrical and Electronic Engineers, Vol. 73, No. 2, 233-251.

Helvey, R.A. 1983. Radiosonde errors and spurious surface-based ducts. Proceedings of the Institution of Electrical Engineers, Vol. 130, Part F, No. 7, 643-648.

Hitney, H.V. 1975. Propagation modelling in the evaporation duct. Naval Electronics Laboratory Center, San Diego, California, NELC/TR 1947.

Hitney, H.V. and J.H. Richter. 1976. Integrated refraction effects prediction system (IREPS). Naval Engineers Journal, April, 257-262.

Hitney, H.V., J.H. Richter, R.A. Pappert, K.D. Anderson, and G.B. Baumgartner, Jr. 1985. Tropospheric radio propagation assessment. Proceedings of the Institute of Electrical and Electronic Engineers, Vol. 73, No. 2, 265-283.

Hoekstra, P. and P. Cappillino. 1971. Dielectric properties of sea and sodium chloride ice at UHF and microwave frequencies. Journal of Geophysical Research, Vol. 76, No. 20, 4922-4931.

Hoekstra, P. and D. Spanogle. 1972. Radar cross-section measurements of snow and ice. U.S. Army Corps of Engineers, Cold Regions Research and Engineering Laboratory, Hanover, N.H., Technical Report 235.

Holt, B., S. Digby, and A. Usan. 1985. Processes and imagery of first-year fast sea ice during the melt season. Journal of Geophysical Research, Vol. 90, No. C3, 5045-5062.

Hood, A.D. 1954a. An analysis of radar ice reports submitted by Hudson Bay Shipping (1953). National Research Council of Canada, Radio and Electrical Engineering Division, Ottawa, No. 3301.

Hood, A.D. 1954b. An analysis of radar ice reports submitted by Hudson Bay Shipping (1954). National Research Council of Canada, Radio and Electrical Engineering Division, Ottawa, No. 3560.

Huynen, R.J. 1965. Measurement of the target scattering matrix. Proceedings of the Institute of Electrical and Electronic Engineers, August, 936-945.

IEE Proceedings - Part A, Special Issue on Historical Radar. 1985. Institution of Electrical Engineers, Vol. 132, Part A, Number 6, October.

Jezek, K.C., J.W. Clough, C.R. Bentley, and S. Shabtaie. 1978. Dielectric permittivity of glacier ice measured in situ by radar wide-angle reflection. Journal of Glaciology, Vol. 21, No. 85, 315-328.

Johnson, J.D. and L.D. Farmer. 1971. Use of side-looking air-borne radar for sea ice identification. Journal of Geophysical Research, Vol. 76, No. 9, 2138-2155.

Katzin, M. 1957. On the mechanisms of radar sea clutter. Proceedings of the Institute of Radio Engineers, January, 44-54.

Kim, Y.-S., R.K. Moore, and R.G. Onstott. 1984. Theoretical and experimental study of radar backscatter from sea ice. Remote Sensing Laboratory, Center for Research, Inc., The University of Kansas, Lawrence, Kansas, Technical Report 331-37.

Kirby, M.E. 1981. Applications of the Roberts gradient operator for the digital enhancement of icebergs from SLAR imagery. The 7th Canadian Remote Sensing Symposium, Winnipeg, Manitoba.

Kirby, M.E. and R.T. Lowry. 1979. Iceberg detectability problems using SAR and SLAR systems. 5th Annual Pecora Symposium: Satellite Hydrology, Sioux Falls, S.D.

Kovacs, A. and R.M. Morey. 1978. Radar anisotropy of sea ice due to preferred azimuthal orientation of the horizontal C axes of the ice crystal. Journal of Geophysical Research, Vol. 83, No. C12, 6037-6046.

Kovacs, A. and R.M. Morey. 1979. Anisotropic properties of sea ice in the 50- to 150-MHz range. Journal of Geophysical Research, Vol. 84, No. C9, 5749-5759.

Kumai, M. and K. Itagaki. 1953. Cinematographic study of ice crystal formation in water. Journal of the Faculty of Science, Hokkaido University, Hokkaido, Japan, Ser. II, 4, 235-246.

Lake, R.A. and E.L. Lewis. 1970. Salt rejection by sea ice during growth. Journal of Geophysical Research, Vol. 75, No. 3, 583-597.

Lamb, J. 1949. The dielectric properties of ice at 1.25 cm wavelength. Proceedings of Physical Society of London, Section B, 272.

Larsson, H. 1949. The use of radar in the ice-breaker service. Journal of the Institute of Navigation, Vol. 2, 315-323.

Le Page, L.S. and A.L.P. Milwright. 1953. Radar and ice. Journal of the Institute of Navigation, Vol. 4, No. 2, 113-130.

Leader, J.C. 1970. Journal of the Optical Society of America, Vol. 60, 1552.

Lewis, B.L. and I.D. Olin. 1980. Experimental study and theoretical model of high-resolution radar backscatter from the sea. Radio Science, Vol. 15, No. 4, 815-828.

Lewis, E.O. and B.W. Currie. 1984a. Effects of frequency and polarization on marine radar detection of ice targets in an ice cover. International Association of Hydraulic Research - Ice Symposium, Hamburg, Germany, 375-384.

Lewis, E.O. and B.W. Currie. 1984b. Improved marine radar display for navigation in ice-covered water. International Association of Hydraulic Research-Ice Symposium, Hamburg, Germany, 363-374.

Lewis, E.O., S. Haykin, and B.W. Currie. 1985. Detection of sea-ice features using K_a-band radar. Electronics Letters, Vol. 21, No. 11, 499-500.

Long, M.W. 1983. Radar reflectivity of land and sea. Artech House, Dedham, Massachusetts.

Luse, J.D. 1981. A brief history of the use of marine radar. Journal of the Institute of Navigation, Vol. 28, No. 3, 199-205.

Maaloe, J. 1982. Sea clutter statistics. Proceedings of the Institutionof Electrical Engineers International Radar Conference, London, England, 193-197.

Marcus, S.W. 1982. A model to calculate EM fields in tropospheric duct environments at frequencies through SHF. Radio Science, Vol. 17, No. 5, 895-901.

Marko, J.R., J.R. Birch, and M.A. Wilson. 1982. A study of long-term satellite-tracked iceberg drifts in Baffin Bay and Davis Strait. Arctic, Vol. 35, No. 1, 234-240.

Martin, S. 1979. A field study of brine drainage and oil entrainment in first-year sea ice. Journal of Glaciology, Vol. 22, No. 88, 473-502.

Mercer, J.B., R.T. Lowry, and S.K. Leving. 1980. Experimental use of real time SAR imagery in support of oil exploration in the Beaufort Sea. 6th Canadian Symposium on Remote Sensing, Halifax, N.S., 143-152.

Moore, J.D. and N.B. Lawrence. 1980. Comparison of two CFAR methods used with square law detection of Swerling I targets. Institute of Electrical and Electronic Engineers International Radar Conference, 403-409.

Morey, R.M., A. Kovacs, and G.F.N. Cox. 1984. Electromagnetic properties of sea ice. U.S. Army Corps of Engineers, Cold Regions Research and Engineering Laboratory, Hanover, N.H., Report 84-2.

Nagao, M. and T. Matsuyama. 1979. Edge preserving smoothing. Computer Graphics and Image Processing, Vol. 9, 394-407.

Nakawo, M. and N.K. Sinha. 1981. Growth rate and salinity profile of first-year sea ice in the high Arctic. Journal of Glaciology, Vol. 27, No. 96, 315-329.

Nakawo, M. 1983. Measurements on air porosity of sea ice. Annals of Glaciology 4, 204-208.

Nathanson, F.E. 1969. Radar design principles-signal processing and the environment. McGraw-Hill, New York, N.Y.

Niedrauer, T.M. and S. Martin. 1979. An experimental study of brine drainage and convection in young sea ice. Journal of Geophysical Research, Vol. 84, No. C3, 1176-1186.

Oi, M., M. Ishikawa, H. Fukushi, and M. Aoto. 1984. Detection of sea ice and ships in radar sea clutter. 1984 International Symposium on Noise and Clutter Rejection in Radars and Imaging Sensors, 141-146.

Onstott, R.G., R.K. Moore, and W.F. Weeks. 1979. Surface-based scatterometer results of Arctic sea ice. Institute of Electrical and Electronic Engineers Transactions on Geoscience Electronics, Vol. GE-17, No. 3, 78-85.

Page, D.F. and R.O. Ramseier. 1975. Application of radar techniques to ice and snow studies. Journal of Glaciology, Vol. 15, No. 73, 171-191.

Page, R.M. 1962. The early history of radar. Proceedings of the Institute of Radio Engineers, May, 1232-1236.

Parashar, S.K. 1974. Investigation of radar discrimination of sea ice. The University of Kansas Center for Research, Inc., Remote Sensing Laboratory, Lawrence, Kansas, Tech. Report 185-13.

Pearce, D.C. and J.W. Walker. 1967. An empirical determination of the relative dielectric constant of the Greenland ice cap. Journal of Geophysical Research, Vol. 72, No. 22, 5743-5747.

Peli, T. and D. Malah. 1982. A study of edge detection algorithms. Computer Graphics and Image Processing, Vol. 20, 1-21.

Perey, F.G.J. and E.R. Pounder. 1958. Crystal formation in ice sheets. Canadian Journal of Physics., Vol. 36, 494-502.

Perry, R.E. 1953. A record of radar performance in ice conditions. Journal of the Institute of Navigation, Vol. Vl, No. 1, 74-85.

Pratt, W.K. 1978. Digital image processing. John Wiley, New York, N.Y.

Richter, J.H. and H.V. Hitney. 1975. The effect of the evaporation duct on microwave propagation. Naval Electronics Laboratory Center, San Diego, California, NELC/TR 1949.

Rohling, H. 1983. Radar CFAR thresholding in clutter and multiple target situations. Institute of Electrical and Electronic Engineers Transactions on Aerospace and Electronic Systems, Vol. AES-19, No. 4, 608-619.

Rossiter, J.R. 1980. Review of impulse radar soundings of sea ice. Proceedings of the International Workshop on Remote Estimation of Sea Ice Thickness, St. John's, Newfoundland, 77-98.

Rotheram, S. 1982. Review of duct propagation. Marconi Research Center, MTR. 82/83.

Rouse, J.W. Jr. 1972. The effect of the subsurface on the depolarization of rough-surface backscatter. Radio Science, Vol. 7, No. 10, 889-895.

Saxton, J.A. and H.G. Hopkins. 1951. Some adverse influences of meteorological factors on marine navigational radar. Proceedings of Institution of Electrical Engineers, Part III, Vol. 98, 26-36.

Scholander, P.F. and D.C. Nutt, D.C. 1960. Bubble pressure in Greenland icebergs. Journal of Glaciology, Vol. 3, No. 28, 671-678.

Schwartz, J. and W.F. Weeks. 1977. Engineering properties of sea ice. Journal of Glaciology, Vol. 19, No. 81, 499-530.

Schwarzacher, W. 1959. Pack-ice studies in the Arctic Ocean. Journal of Geophysical Research, Vol. 64, No. 12, 2357-2367.

Sekine, M., T. Musha, Y. Tomita, T. Hagisawa, and T. Irabu. 1980. Suppression of Weibull-distributed weather clutter. Institute of Electrical and Electronic Engineers International Radar Conference, 294-298.

Shaw, G.B. 1979. Local and regional edge detectors: some comparisons. Computer Graphics and Image Processing, Vol. 9, 135-149.

Sheives, T.C. 1974. A study of a dual polarization laser backscatter system for remote identification and measurement of water pollution. Remote Sensing Center, Texas A&M University, College Station, Texas, Technical Report RSC-52, 131 p.

Sinha, N.K. 1985. Verbal communication. National Research Council of Canada, Building Research Division, Ottawa.

Sittrop, H. 1977. On the sea-clutter dependency on windspeed. Institution of Electrical Engineers International Radar Conference, London, England, 110-114.

Skolnik, M.I. 1962. Introduction to radar systems. McGraw-Hill, New York, N.Y.

Skolnik, M.I. 1970. Radar handbook. McGraw-Hill, New York, N.Y.

Skolnik, M.I. 1980. Introduction to radar systems. McGraw-Hill, New York, N.Y.

Skolnik, M.I. 1985. Fifty years of radar. Proceedings of the Institute of Electrical and Electronic Engineers, Vol. 73, No. 2, 182-190.

Smith, J.M. 1984. Fast-scan processing in maritime surveillance radars. 1984 International Symposium on Noise and Clutter Rejection in Radars and Imaging Sensors, 506-511.

Swerling, P. 1954. Probability of detection for fluctuating targets. RAND Corp. Research Memo, RM-1217. [Reprinted in Transactions of the Institute of Radio Engineers, IT-6, No. 2, April 1960.]

Tabata, T. 1975. Sea-ice reconnaissance by radar. Journal of Glaciology, Vol. 15, No. 73, 215-224.

Tabata, T., M. Aota, and and M. Oi. 1967. On the observation of sea ice distribution with the sea ice radar (preliminary report). Low Temperature Science, Series A (Physical Sciences), No. 25, 233-239. Translated by E.R. Hope, 1970, Defence Research Board of Canada, Ottawa, T102J.

Tabata, T., M. Aota, M. Oi, and M. Ishikawa. 1969a. Observation of drift ice movements with the sea ice radar network. Low Temperature Science, Series A (Physical Sciences), No. 17, 295-315. Translated by E.R. Hope, 1970, Defence Research Board of Canada, Ottawa, T105J.

Tabata, T., M. Aota, M. Oi, and M. Ishikawa. 1969b. Distribution of pack ice field off Hokaido observed with sea ice radar network. Low Temperature Science, Series A (Physical Sciences), No. 27, 23-38, Data Report Supplement. Translated by E.R. Hope, Defence Research Board of Canada, Ottawa, T103J.

Ulaby, F.T., R.K. Moore, and A.K. Fung. 1982. Remote sensing - active and passive, Volume II. Addison-Wesley, Reading, Massachusetts.

Vant, M.R., R.B. Gray, R.O. Ramseier, and V. Makios. 1974. Dielectric properties of fresh and sea ice at 10 and 35 GHz. Journal of Applied Physics, Vol. 45, No. 11, 4712-4717.

Vant, M.R., R.O. Ramseier, and V. Makios. 1978. The complex-dielectric constant of sea ice at frequencies in the range 0.1-40 GHz. Journal of Applied Physics, Vol. 49, No. 3, 1264-1280.

Von Hippel, A. 1954. Tables of dielectric materials. Massachusetts Institute of Technology Press.

Walker, E.R. and P. Wadhams. 1979. Thick sea-ice floes. Arctic, Vol. 32, No. 2, 140-147.

Weeks, W.F. and O.S. Lee. 1958. Observations on the physical properties of sea ice at Hopedale, Labrador. Arctic, Vol. 11, 135-155.

Weeks, W.F. and O.S. Lee. 1962. The salinity distribution in young sea ice. Arctic, Vol. 15, 93-108.

Weeks, W.F. and S.F. Ackley. 1982. The growth, structure and properties of sea ice. U.S. Army Corps of Engineers Cold Regions Research and Engineering Laboratory, Hanover, N.H., Monograph 82-1.

Williams, P.D.L. 1973. Detection of sea ice growlers by radar. Institute of Electrical and Electronic Engineers Radar Conference, 239-244.

Williams, P.D.L. 1975. Limitations of radar techniques for the detection of small surface targets in clutter. Radio and Electronic Engineer, Vol. 45, No. 8, 379-389.

Williams, P.D.L. 1979. The detection of ice at sea by radar. The Radio and Electronic Engineer, Vol. 49, No. 6, 275-287.

Wilson, J.T., J.H. Zumberge, and E.W. Marshall. 1954. A study of ice on an inland lake. U.S.A. Snow, Ice and Permafrost Research Establishment, Research Report No. 5.

Yamada, Y., H. Yanagisawa, and Y. Ohhashi. 1984. Sea clutter analysis for adaptive CFAR design. 1984 International Symposium on Noise and Clutter Rejection in Radars and Imaging Sensors, IECE, 117-122.

Yang, G.J. and T.S. Huang. 1985. The effect of median filtering on edge location estimation. Computer Graphics and Image Processing, Vol. 15, 224-245.

APPENDIX A
Ice Nomenclature

Adapted from World Meteorological Organization, Sea Ice Nomenclature, WMO No. 259.TP.145

Aged Ridge Ridge which has undergone considerable weathering. These ridges are best described as undulations.

Bergy Bit A large piece of floating glacier ice, generally showing less than 5 m above sea-level but more than 1m and normally about 100 to 300 m² in area.

Calving The breaking away of a mass of ice from an ice wall, ice front, or iceberg.

Concentration The ratio in tenths of the sea surface actually covered by ice to the total area of sea surface, both ice-covered and ice-free, at a specific location or over a defined area. Total concentration includes all stages of development that are present.

Consolidated Ice Floating ice in which the concentration is 10/10 and the floes are frozen together.

Consolidated Ridge A ridge in which the base has frozen together.

Deformed Ice A general term for ice which has been squeezed together and in places forced upwards (and downwards). Subdivisions are rafted ice, ridged ice, and hummocked ice.

Drift Ice/Pack Ice Term used in a wide sense to include any area of sea ice other than fast ice no matter what form it takes or how it is dispersed. When concentrations are high (i.e., 7/10 or more), drift ice may be replaced by the term pack ice. Previously the term pack ice was used for all ranges of concentration.

Fast Ice Sea ice which forms and remains fast along the coast, where it is attached to the shore, to an ice wall, to an ice front, between shoals or grounded icebergs. Vertical fluctuations may be observed during changes of sea level. Fast ice may be formed *in situ* from sea water or by freezing of pack ice of any age to the shore, and it may extend a few metres, or several hundred kilometres from the coast. Fast ice may be more than one year old and may then be prefixed with the appropriate age category (old, second-year, or multi-year). If it is thicker than about 2m above sea level it is called an ice shelf.

Fast-Ice Edge The demarcation at any given time between fast ice and open water.

Firn Old snow which has recrystallized into a dense material. Unlike snow, the particles are to some extent joined together; but, unlike ice, the air spaces in it still connect with each other.

First-Year Ice Sea ice of not more than one winter's growth, developing from young ice; thickness 30 cm - 2 m. May be subdivided into thin first-year ice/white ice, medium first-year ice, and thick first-year ice.

Floe Any relatively flat piece of sea ice 20 m or more across. Floes are subdivided according to horizontal extent as follows:

Giant:	Over 10 km across
Vast:	2 - 10 km across
Big:	500 - 2,000 m across
Medium:	100 - 500 m across
Small:	20 - 100 m across.

Flooded Ice Sea ice which has been flooded by melt-water or river water and is heavily loaded by water and wet snow.

Fracture Any break or rupture through very close pack ice, compact pack ice, consolidated pack-ice, fast ice, or a single floe resulting from deformation processes. Fractures may contain brash ice and/or be covered with nilas and/or young ice. Length may vary from a few metres to many kilometres.

Crack:	0 - 1 m wide
Very Small Fracture:	1 - 50 m wide
Small Fracture:	50 - 200 m wide
Medium Fracture:	200 - 500 m wide
Large Fracture:	More than 500 m wide.

Frazil Ice Fine spicules or plates of ice, suspended in water.

Glacier A mass of snow and ice continuously moving from higher to lower ground or, if afloat, continuously spreading. The principal forms of glacier are: inland ice sheets, ice shelves, ice streams, ice caps, ice pied-monts, cirque glaciers, and various types of mountain (valley) glaciers.

Glacier Berg An irregularly shaped iceberg.

Glacier Ice Ice in, or originating from, a glacier, whether on land or floating on the sea as icebergs, bergy bits, or growlers.

Glacier Tongue Projecting seaward extension of a glacier, usually afloat. In the Antarctic glacier tongues may extend over many tens of kilometres.

Grease Ice A later stage of freezing than frazil ice when the crystals have coagulated to form a soupy layer on the surface. Grease ice reflects little light, giving the sea a matt appearance.

Grey Ice Young ice 10 - 15 cm thick. Less elastic than nilas and breaks on swell. Usually rafts under pressure.

<u>Grey-White Ice</u> Young ice 15 - 30 cm thick. Under pressure more likely to ridge than to raft.

<u>Grounded Ice</u> Floating ice which is aground in shoal water.

<u>Growler</u> Smaller piece of ice than a bergy bit or floe-berg, often transparent but appearing green or almost black in colour, extending less than 1 m above the sea surface and normally occupying an area of about 20 m^2.

<u>Hummock</u> A hillock of broken ice which has been forced upwards by pressure. May be fresh or weathered. The submerged volume of broken ice under the hummock, forced downwards by pressure, is termed a bummock.

<u>Iceberg</u> A massive piece of ice of greatly varying shape, more than 5 m above sea level, which has broken away from a glacier, and which may be afloat or aground. Icebergs may be described as tabular, dome-shaped, sloping, pinnacled, weathered, or glacier bergs.

<u>Ice Breccia</u> Ice pieces of different age frozen together.

<u>Ice Cover</u> The ratio of an area of ice of any concentration to the total area of sea surface within some large geographic locale; this locale may be global, hemispheric, or prescribed by a specific oceanographic entity such as Baffin Bay or the Barents Sea.

<u>Ice Edge</u> The demarcation at any given time between the open sea and sea ice of any kind, whether fast or drifting. It may be termed compacted or diffuse.

<u>Ice Field</u> Area of floating ice consisting of any size of floes, which is greater than 10 km across.

<u>Ice-Free</u> No sea ice present.

<u>Ice Island</u> A large piece of floating ice about 5 m above sea level, which has broken away from an Arctic ice shelf, having a thickness of 30 - 50 m and an area of from a few thousand square metres to 500 km^2 or more, and

usually characterized by a regularly undulating surface which gives it a ribbed appearance from the air.

Ice Keel A downward-projecting ridge of the underside of the ice canopy; the counterpart of a ridge. Ice keels may extend as much as 50 m below sea level.

Ice Shelf A floating ice sheet of considerable thickness showing 2 - 50 m or more above sea level, attached to the coast. Usually of great horizontal extent and with a level or gently undulating surface. Nourished by annual snow accumulation and often also by the seaward extension of land glaciers. Limited areas may be aground. The seaward edge is termed an ice front.

Lead Any fracture or passageway through sea ice which is navigable by surface vessels.

Level Ice Sea ice which is unaffected by deformation.

Multiyear Ice Old ice up to 3 m or more thick which has survived at least two summers' melt. Hummocks even smoother than in second-year ice, and the ice is almost salt-free. Colour, where bare, is usually blue. Melt pattern consists of large, interconnecting irregular puddles and a well-developed drainage system.

New Ice A general term for recently formed ice which includes frazil ice, grease ice, slush, and shuga. These types of ice are composed of ice crystals which are only weakly frozen together (if at all) and have a definite form only while they are afloat.

New Ridge Ridge newly formed with sharp peaks and slope of sides usually 40°. Fragments are visible from the air at low altitude.

Nilas A thin elastic crust of ice, easily bending on waves and swell and under pressure, thrusting in a pattern of interlocking "fingers" (finger rafting). Has a matt surface and is up to 10 cm in thickness. May be subdivided into dark nilas and light nilas.

<u>Old Ice</u> Sea ice which has survived at least one summer's melt. Most topographic features are smoother than on first-year ice. May be subdivided into second-year ice and multiyear ice.

<u>Open Water</u> A large area of freely navigable water in which sea ice is present in concentrations less than 1/10. When there is no sea ice present, the area should be termed ice-free, even though icebergs are present.

<u>Pack Ice</u> Term used in a wide sense to include any area of sea ice, other than fast ice, no matter what form it takes or how it is disposed.

<u>Pancake Ice</u> Predominantly circular pieces of ice from 30 cm - 3 m in diameter, and up to about 10 cm in thickness, with raised rims due to the pieces striking against one another. It may be formed on a slight swell from grease ice, shuga, or slush or as a result of the breaking of ice rind, nilas or, under severe conditions of swell or waves, of grey ice. It also sometimes forms at some depth, at an interface between water bodies of different physical characteristics, from where it floats to the surface; its appearance may rapidly cover wide areas of water.

<u>Polynya</u> Any non-linear shaped opening enclosed in ice. Polynyas may contain brash ice and/or be covered with new ice, nilas, or young ice; submariners refer to these as skylights. Sometimes the polynya is limited on one side by the coast and is called a shore polynya or by fast ice and is called a flaw polynya. If it recurs in the same position every year, it is called a recurring polynya.

<u>Puddle</u> An accumulation on ice of melt-water, mainly due to melting snow, but in the more advanced stages also to the melting of ice. Initial stage consists of patches of melted snow.

Rafted Ice Type of deformed ice formed by one piece of ice overriding another.

Rafting Pressure processes whereby one piece of ice overrides another.

Ridge A line or wall of broken ice forced up by pressure. May be fresh or weathered. The submerged volume of broken ice under a ridge, forced downwards by pressure, is termed an ice keel.

Ridged Ice Ice piled haphazardly one piece over another in the form of ridges or walls. Usually found in first-year ice (cf. ridging).

Ridging The pressure process by which sea ice is forced into ridges.

Sea Ice Any form of ice found at sea which has originated from the freezing of sea water.

Second-Year Ice Old ice which has survived only one summer's melt. Because it is thicker and less dense than first-year ice, it stands higher out of the water. In contrast to multiyear ice, summer melting produces a regular pattern of numerous small puddles. Bare patches and puddles are usually greenish-blue.

Tabular Berg A flat-topped iceberg. Most tabular bergs form by calving from an ice shelf and show horizontal banding (cf. ice island).

Thick First-Year Ice First-year ice over 120 cm thick.

Thin First-Year Ice/White Ice First-year ice 30 - 70 cm thick.

Weathered Ridge Ridge with peaks slightly rounded and slope of sides usually 30 - 40°. Individual fragments are not discernible.

Young Ice Ice in the transition stage between nilas and first-year ice, 10 - 30 cm in thickness. May be subdivided into grey ice and grey-white ice.

APPENDIX B
Equipment Specifications

Furuno FRM 64W

Frequency	• X-band 9.380 GHz (measured)
Transmitter	
Peak power	• 8.8 kw
Pulse length	• 1,000 ns, 100 ns
PRF	• 799 Hz, 3194 Hz (measured)
Receiver	
Type	• logarithmic
IF	• 60 MHz
Noise figure	• 10 dB
MDS	• 96.5 dBm
Antenna	
Type	• slotted waveguide
Length	• 6.5 ft (2 m)
Gain	• 29.5 dB (specified), 31 dB (measured)
Horizontal beamwidth	• 1.2° (3 dB)
Vertical beamwidth	• 25° (3 dB)
Rotation rate	• 24 rpm
Polarization	• horizontal
Antenna	
Type	• parabolic dish
Diameter	• 1 m
Gain	• 36 dB
Beam pattern	• pencil, 1.2°

Decca RM 1229

Frequency	• X-band 9.380 GHz
Transmitter	
Peak power	• 25 kw, 24.4 kw (measured)
Pulse length	• 1,000 ns, 250 ns, 50 ns
PRF	• 834 Hz, 1,660 Hz, 3,320 Hz (measured)
Receiver	
Type	• logarithmic
IF	• 60 MHz
Noise figure	• 10 dB
MDS	• 96 dBm
IF bandwidth	• 18 MHz (50 ns pulse length) 5 MHz (1,000 ns, 250 ns pulse length)
Video bandwidth	• 16 MHz nominal

Antenna	
Type	• slotted waveguide
Length	• 4 ft (1.2 m) 9 ft (2.1 m)
Gain	• 28 dB (24 dB measured) 33 dB (31 dB measured)
Horizontal beamwidth	• 1.9° (3 dB) 0.8° (3 dB)
Vertical beamwidth	• 20° (3 dB)
Rotation rate	• 28 rpm
Polarization	• horizontal

Decca RMS 1230

Frequency	• S-band 3.040 GHz
Transmitter	
Peak power	• 30 kw (30.8 kw measured)
Pulse length	• 1,000 ns, 250 ns, 25 ns
PRF	• 829 Hz, 1,654 Hz, 3,320 Hz (measured)
Receiver	
Type	• logarithmic
IF	• 60 MHz
Noise figure	• 10 dB
MDS	• 95 dBm
IF bandwidth	• 18 MHz (50 ns pulse length) 5 MHz (1,000 ns, 250 ns pulse length)
Video bandwidth	• 16 MHz nominal
Antenna	
Type	• slotted waveguide
Length	• 12 ft (3.7 m)
Gain	• 27 dB
Horizontal beamwidth	• 2° (3 dB)
Vertical beamwidth	• 25° (3 dB)
Rotation rate	• 22 rpm
Polarization	• horizontal

RCA Adviser Recorder

Power requirement	115 V rms, 400 Hz - 60 Hz three phase, single phase, 650 W
Number of channels	a) 2 wideband video channels b) auxiliary channel 1 c) auxiliary channel 2
Record time and tape speed	a) dual channel - 23 minutes at 15 ips —— {8 in reel configuration b) single channel - 46 minutes at 7.5 ips dual channel - 46 minutes at 15 ips —— {1½ in reel configuration single channel - 90 minutes at 7.5 ips
Wide band video channel characteristics	
Frequency response	15.0 Hz - 5.0 MHz ± 3 dB at 6 MHz and 10 Hz
Input/output impedance	75 ohms
Signal-to-noise ratio	36 dB peak-to-peak rms signal in reference to 1.0 V peak-to-peak signal between 10 Hz and 6 MHz
Rise/fall time	120 ns maximum
Overshoot	±10% maximum
Amplitude linearity	±0.5 dB from 1.0 V peak-to-peak to 0.1 V peak-to-peak
Amplitude stability	±0.5 dB at 1.0 V peak-to-peak
Tape format	ASA standard C98, 6-1965 (half-track transverse scan)
Output level	1.0 V peak-to-peak (adjustable)
Time-base stability	75 ns rms absolute
Head switching transients	6 dB min below peak noise
Crosstalk	not greater than 35 dB below reference signal of 1.0 V peak-to-peak

Auxiliary channel 1

Characteristics

Frequency response	100 Hz to 15 kHz ± 3 dB
Input/output impedance (voice)	600 ohms
Signal-to-noise ratio	30 dB peak-to-peak rms signal
Input/output level	1.0 V rms
Flutter	0.4% rms flutter from 0.6 Hz - 200 Hz
Harmonic distortion	less than 1.0% using a 3 kHz, 1.0 V rms reference signal (third harmonic)

Reel size	NAB standard 8 or 10.5 in
Erase	capability to erase all channels

Operating condition

Temperature	-30 to +55°C
Operating altitude	0 - 40,000 ft ASL
Humidity	to 95% RH
Vibration	MIL-E-5400, Curve II

Litton/Decca X-Band Radar Combination

Transmitter - Litton APS-504 (V3)

Type	• magnetron
Frequency	• 9.4 Hz
Peak power	• 100 kw
Pulse width	• 510 μs narrow pulse • 2.1 μs wide pulse
PRF	• 1,668 Hz narrow pulse, short range • 212 Hz wide pulse, long range

Dual-polarized antenna

Type	• paraboloid
Diameter	• 30 in (0.8 m)
Vertical beamwidth	• 3.5°
Horizontal beamwidth	• 3.5°
Gain	• 33 dB (HH and VV)
Polarization	• horizontal and vertical
Waveguide losses (one way)	• 3 dB cross (Decca)

Receivers	Litton	Decca (Receive Only)
Type	• Superheterodyne	• Logarithmic amplitude
Detection	• Horizontal amplitude	• Vertical amplitude
IF frequency	• 60 MHz	• 60 MHz
Bandwidth	• 18 MHz, 5 MHz	• 18 MHz, 5 MHz
MDS	• -103 dBm	• -96 dBm

Coherent X-Band Radar

Transmitter

Type	• master oscillator with travelling wave tube amplifier
RF frequency	• 9.4 GHz (nominal)
Peak power	• 1 kw (nominal)
Pulse width	• 200 ns - narrow pulse
PRF	• 0 - 2 kHz

Receiver

Type	• logarithmic amplitude, vertical and horizontal • linear vertical I and Q • linear horizontal I and Q
Beamwidth	• 10 MHz logarithmic • 10 MHz linear
Dynamic range	• 70 dB logarithmic • 35 dB linear
Sensitivity	• -90 dBm

Dual-polarized antenna

Type	• paraboloid
Diameter	• 30 in (0.8 m)
Vertical beamwidth	• 3.5°
Horizontal beamwidth	• 3.5°
Gain	• 33 dB (HH and VV)
Polarization	• vertical and horizontal

Calibration

Injected signal	• 20 dB coupler in vertical and horizontal waveguide paths
Reference target	• 50 dB RF attenuator in vertical and horizontal waveguide paths

K_u-Band Radar

Transmitter

Type	• magnetron
RF frequency	• 16.5 GHz
Peak power	• 28 kw
Pulse width	• 400 ns
PRF	• 1,538 Hz

Dual-polarized antenna

Type	• paraboloid
Diameter	• 3 ft (0.9 m)
Vertical beamwidth	• 1.5° (3 dB points)
Horizontal beamwidth	• 1.5° (3 dB points)
Gain	• 41.5 dB (HH and VV)
Polarization	• vertical and horizontal
Waveguide loss	• 3.5 dB vertical, 4.5 dB horizontal

Receivers, horizontal and vertical

Type	• superheterodyne, log IF amplifiers
Detection	• amplitude
Bandwidth	• 20 MHz, log IF
Dynamic range	• 70 dB
MDS	• -73 dBm (parallel)
	• -68 dBm (cross)

K_a-Band Radar

Transmitter

Type	• magnetron
RF frequency	• 35.0 GHz
Peak power	• 4.5 kw
Pulse width	• 140 ns
PRF	• 1 kHz

Dual-polarized antenna

Type	• paraboloid
Diameter	• 2 ft (0.6 m)
Vertical beamwidth	• 1.0° (3 dB points)
Horizontal beamwidth	• 1.0° (3 dB points)
Gain	• 43 dB (HH and VV)
Polarization	• vertical and horizontal
Waveguide loss (one way)	• 15 dB vertical, 15 dB horizontal

Receivers, horizontal and vertical

Type	• superheterodyne, log IF amplifiers
Detection	• vertical and horizontal amplitude
IF frequency	• 60 MHz
Bandwidth	• 20 MHz
Dynamic range	• 70 dB
MDS	• -67 dBm (parallel) • -69 dBm (cross)

Digital Data Reduction Facility

Parameter	Description
Central processor	NOVA-2 16-bit minicomputer with 64 K bytes memory, floating point arithmetic unit
A/D converter	12-bit, 16 channels multiplex, 20-K word/s sample date
Peripherals	10 M byte disc, plotting CRT terminal, IDS-560 line printer with graphics capability, dual digital tape drives
Data reduction routines	Calibration, probability densities, cumulative distributions, average values, standard deviations, autocorrelation functions, power spectral densities, range profiles, waterfall plots

Data outputs	Magnetic analogue and digital tapes, printed copy, data plots
Data inputs	Analogue signals, digital tapes, wideband video signals
Computer controlled	1,024 range positions, 80 ns range resolution

Furuno FR-1011

Frequency	• X-band 9.410 GHz
Transmitter	
Peak power	• 10 kw • 7 kw (medium pulse)
Pulse length	• 1.0 µs, 600 ns, 80 ns
PRF	• 3,200 Hz, 1,600 Hz, 800 Hz, 600 Hz
Receiver	
Type	• logarithmic
IF	• 40 MHz
MDS	• 96 dBm
Antenna	
Type	• parabolic dish
Diameter	• 1.2 m
Gain	• 38.5 dB
Beam pattern	• 2.2°
Rotation rate	• 24 rpm
Polarization	• horizontal and vertical

Video Data Acquisition System (VDAS)

Digitizer	
Video digitizing rate	• 5 megasamples/s
Resolution	• 8 bits
Channels	• 2
Digitizing format	• 128 scans of 64 ranges • 16 scans of 512 ranges
Antenna modes	• fixed or rotating
Range delay	• 0 - 4095 µs
Start angle	• 0 - 359 degrees
Controller/Logger	IBM PCXT with 10 megabyte hard disk

Index